P9-BYG-531

TIME TRAVEL

ALSO BY JAMES GLEICK

Chaos:
Making a New Science

Genius:
The Life and Science of Richard Feynman

Faster:
The Acceleration of Just About Everything

What Just Happened:
A Chronicle from the Information Frontier

Isaac Newton

The Information:
A History, a Theory, a Flood

TIME TRAVEL

A History

JAMES GLEICK

4th ESTATE • *London*

4th Estate
An imprint of HarperCollins*Publishers*
1 London Bridge Street
London SE1 9GF
www.4thEstate.co.uk

First published in Great Britain by 4th Estate in 2016
First published in the United States by Pantheon Books,
a division of Penguin Random House LLC, in 2016

This edition 2017

1 3 5 7 9 10 8 6 4 2

Copyright © James Gleick 2016

James Gleick asserts the moral right to be identified as the author of this work

Grateful acknowledgment is made to Houghton Mifflin Harcourt Publishing Company
for permission to reprint excerpts from "Burnt Norton" and "The Dry Salvages" from
Four Quartets by T. S. Eliot, copyright © 1936 by Houghton Mifflin Harcourt Publishing Company,
renewed 1964 by T. S. Eliot, and renewed 1969 by Esme Valerie Eliot. Reprinted by permission
of Houghton Mifflin Harcourt Publishing Company. All rights reserved.

Grateful acknowledgement is also made to Faber & Faber Ltd for permission to reprint
excerpts from "Burnt Norton" and "The Dry Salvages" from *Four Quartets* by T.S. Eliot.
© The Estate of T.S. Eliot. Reproduced by kind permission of Faber & Faber Ltd.

A catalogue record of this book is available from the British Library

Hardback ISBN 978-0-00-754443-1
Trade paperback ISBN 978-0-00-820767-0

Designed by Cassandra J. Pappas

Printed and bound in Great Britain by Clays Ltd, St Ives plc

All rights reserved. No part of this publication may be
reproduced, stored in a retrieval system, or transmitted,
in any form or by any means, electronic, mechanical,
photocopying, recording or otherwise, without the
prior written permission of the publishers.

This book is sold subject to the condition that it shall not, by
way of trade or otherwise, be lent, re-sold, hired out of otherwise
circulated without the publisher's prior consent in any form of
binding or cover other than that in which it is published and
without a similar condition including this condition being
imposed on the subsequent purchaser.

FSC™ is a non-profit international organisation established to promote
the responsible management of the world's forests. Products carrying the
FSC label are independently certified to assure consumers that they come
from forests that are managed to meet the social, economic and
ecological needs of present and future generations,
and other controlled sources.

Find out more about HarperCollins and the environment at
www.harpercollins.co.uk/green

To Beth, Donen,
and Harry

Your now is not my now; and again, your then is not my then; but my now may be your then, and vice versa. Whose head is competent to these things?

—Charles Lamb (1817)

The fact that we occupy an ever larger place in Time is something that everybody feels.

—Marcel Proust (1927?)

And tomorrow
Comes. It's a world. It's a way.
—W. H. Auden (1936)

Contents

ONE

Machine

Being young, I was skeptical of the future, and saw it as a matter of potential only, a state of things that might or might not arise and probably never would.

—John Banville (2012)

A MAN STANDS AT the end of a drafty corridor, a.k.a. the nineteenth century, and in the flickering light of an oil lamp examines a machine made of nickel and ivory, with brass rails and quartz rods—a squat, ugly contraption, somehow out of focus, not easy for the poor reader to visualize, despite the listing of parts and materials. Our hero fiddles with some screws, adds a drop of oil, and plants himself on the saddle. He grasps a lever with both hands. He is going on a journey. And by the way so are we. When he throws that lever, time breaks from its moorings.

The man is nondescript, almost devoid of features—"grey eyes" and a "pale face" and not much else. He lacks even a name. He is just the Time Traveller: "for so it will be convenient to speak of him." *Time* and *travel:* no one had thought to join those words before now. And that machine? With its saddle and bars, it's a fantasticated bicycle. The whole thing is the invention of a young enthusiast named Wells, who goes by his initials, H. G., because he thinks that sounds more serious than Herbert. His family calls him Bertie. He is trying to be a writer. He is a thoroughly modern man, a believer in socialism, free love, and bicycles.* A proud member of the Cyclists' Touring Club, he rides up and down the Thames valley on a forty-pounder with tubular frame and pneumatic tires, savoring the thrill of riding his machine: "A memory of motion lingers in the muscles of your legs, and round and round they seem to go." At some point he sees a printed advertisement for a contraption called Hacker's Home Bicycle: a stationary stand with rubber wheels to let a person pedal for exercise without going anywhere. Anywhere through space, that is. The wheels go round and time goes by.

The turn of the twentieth century loomed—a calendar date with apocalyptic resonance. Albert Einstein was a boy at gymnasium in Munich. Not till 1908 would the Polish-German mathematician Hermann Minkowski announce his radical idea: "Henceforth space by itself, and time by itself, are doomed to fade away into mere shadows, and only a kind of union of the two will preserve an independent reality." H. G. Wells was there first, but unlike Minkowski, Wells was not trying to explain the universe. He was just trying to gin up a plausible-sounding plot device for a piece of fantastic storytelling.

Nowadays we voyage through time so easily and so well, in our dreams and in our art. Time travel feels like an ancient tradition, rooted in old mythologies, old as gods and dragons. It isn't. Though the ancients imagined immortality and rebirth and lands of the dead

* He defined free love as "the liberation of individual sexual conduct from social reproach and from legal controls and penalties." And he "practised it tirelessly," as David Lodge wrote.

time machines were beyond their ken. Time travel is a fantasy of the modern era. When Wells in his lamp-lit room imagined a time machine, he also invented a new mode of thought.

Why not before? And why now?

THE TIME TRAVELLER BEGINS with a science lesson. Or is it just flummery? He gathers his friends around the drawing-room fire to explain that everything they know about time is wrong. They are stock characters from central casting: the Medical Man, the Psychologist, the Editor, the Journalist, the Silent Man, the Very Young Man, and the Provincial Mayor, plus everyone's favorite straight man, "an argumentative person with red hair" named Filby.

"You must follow me carefully," the Time Traveller instructs these stick figures. "I shall have to controvert one or two ideas that are almost universally accepted. The geometry, for instance, that they taught you at school is founded on a misconception." School geometry—Euclid's geometry—had three dimensions, the ones we can see: length, width, and height.

Naturally they are dubious. The Time Traveller proceeds Socratically. He batters them with logic. They put up feeble resistance.

> "You know of course that a mathematical line, a line of thickness *nil,* has no real existence. They taught you that? Neither has a mathematical plane. These things are mere abstractions."
>
> "That is all right," said the Psychologist.
>
> "Nor, having only length, breadth, and thickness, can a cube have a real existence."

"There I object," said Filby. "Of course a solid body may exist. All real things—"

"So most people think. But wait a moment. Can an *instantaneous* cube exist?"

"Don't follow you," said Filby [the poor sap].

"Can a cube that does not last for any time at all, have a real existence?"

Filby became pensive. "Clearly," the Time Traveller proceeded, "any real body must have extension in *four* directions: it must have Length, Breadth, Thickness, and—Duration."

Aha! The fourth dimension. A few clever Continental mathematicians were already talking as though Euclid's three dimensions were not the be-all and end-all. There was August Möbius, whose famous "strip" was a two-dimensional surface making a twist through the third dimension, and Felix Klein, whose loopy "bottle" implied a fourth; there were Gauss and Riemann and Lobachevsky, all thinking, as it were, outside the box. For geometers the fourth dimension was an unknown direction at right angles to all our known directions. Can anyone visualize that? What direction is it? Even in the seventeenth century, the English mathematician John Wallis, recognizing the algebraic possibility of higher dimensions, called them "a Monster in Nature, less possible than a Chimaera or Centaure." More and more, though, mathematics found use for concepts that lacked physical meaning. They could play their parts in an abstract world without necessarily describing features of reality.

Under the influence of these geometers, a schoolmaster named Edwin Abbott Abbott published his whimsical little novel *Flatland: A Romance of Many Dimensions* in 1884, in which two-dimensional creatures try to wrap their minds around the possibility of a third; and in 1888 Charles Howard Hinton, a son-in-law of the logician George Boole, invented the word *tesseract* for the four-dimensional analogue of the cube. The four-dimensional space this object encloses he called hypervolume. He populated it with hypercones, hyperpyramids, and hyperspheres. Hinton titled his book, not very modestly, *A New Era of Thought.* He suggested that this mysterious, not-quite-visible fourth dimension might provide an answer to the mystery of consciousness. "We must be really four-dimensional creatures, or we could not think about four dimensions," he reasoned. To make mental models of the world and of ourselves, we must have special brain molecules: "It may be that these brain molecules have the power of four-dimensional movement, and that they can go through four-dimensional movements and form four-dimensional structures."

For a while in Victorian England the fourth dimension served as a catchall, a hideaway for the mysterious, the unseen, the spiritual—anything that seemed to be lurking just out of sight. Heaven might be in the fourth dimension; after all, astronomers with their telescopes were not finding it overhead. The fourth dimension was a secret compartment for fantasists and occultists. "We are on the eve of the Fourth Dimension; that is what it is!" declared William T. Stead, a muckraking journalist who had been editor of the *Pall Mall Gazette,* in 1893. He explained that this could be expressed by mathematical formulas and could be imagined ("if you have a vivid imagination") but could not actually be seen—anyway not "by mortal man." It was a place "of which we catch glimpses now and then in those phenomena which are entirely unaccountable for by any law of three-dimensional space." For example, clairvoyance. Also telepathy. He submitted his report to the Psychical Research Society for their further investigation. Nineteen years later he embarked on the *Titanic* and drowned at sea.

By comparison Wells is so sober, so simple. No mysticism for him—the fourth dimension is not a ghost world. It is not heaven, nor is it hell. It is time.

What is time? Time is nothing but one more direction, orthogonal to the rest. As simple as that. It's just that no one has been able to see it till now—till the Time Traveller. "Through a natural infirmity of the flesh . . . we incline to overlook this fact," he coolly explains. *"There is no difference between Time and any of the three dimensions of Space except that our consciousness moves along it."*

In surprisingly short order this notion would become part of the orthodoxy of theoretical physics.

WHERE DID THE IDEA come from? There was something in the air. Much later Wells tried to remember:

> In the universe in which my brain was living in 1879, there was no nonsense about time being space or anything of that sort. There were three dimensions, up and down, fore and aft and right and left, and I never heard of a fourth dimension until 1884 or thereabout. Then I thought it was a witticism.

Very witty. People of the nineteenth century sometimes asked, as people will, "What is time?" The question arises in many different contexts. Say you want to explain the Bible to children. The *Educational Magazine,* 1835:

> Ver. 1. In the beginning God created the heaven and the earth.
>
> What do you mean by the beginning? *The beginning of time.*— What is time? *A measured portion of eternity.*

But everyone knows what time is. It was true then and it's true now. Also no one knows what time is. Augustine stated this pseudo-paradox in the fourth century and people have been quoting him, wittingly and unwittingly, ever since:

> What then is time? If no one asks me, I know. If I wish to explain it to one that asks, I know not.*

Isaac Newton said at the outset of the *Principia* that everyone knew what time was, but he proceeded to alter what everyone knew. Sean Carroll, a modern physicist, says (tongue in cheek), "Everybody knows what time is. It's what you find out by looking at a clock." He also says, "Time is the label we stick on different moments in the life of the world." Physicists like this bumper-sticker game. John Archibald Wheeler is supposed to have said, "Time is nature's way to keep everything from happening all at once," but Woody Allen said that, too, and Wheeler admitted having found it scrawled in a Texas men's room.†

* *Quid est ergo tempus? Si nemo ex me quaerat, scio; si quaerenti explicare velim, nescio.*

† Predating them by several decades, a science-fiction writer named Ray Cummings put those words into the mouth of a character called the Big Business Man in his 1922 novel *The Girl in the Golden Atom.* Later, Susan Sontag said (quoting "an old riff I've always imagined to have been invented by some graduate student of philosophy"), "Time exists in order that everything doesn't happen all at once, and space exists so that it doesn't all happen to you."

Richard Feynman said, "Time is what happens when nothing else happens," which he knew was a wisecrack. "Maybe it is just as well if we face the fact that time is one of the things we probably cannot define (in the dictionary sense), and just say that it is what we already know it to be: it is how long we wait."

When Augustine contemplated time, one thing he knew was that it was not space—"and yet, Lord, we perceive intervals of times, and compare them, and say some are shorter, and others longer." We *measure* time, he said, though he had no clocks. "We measure times as they are passing, by perceiving them; but past, which now are not, or the future, which are not yet, who can measure?" You cannot measure what does not yet exist, Augustine felt, nor what has passed away.

In many cultures—but not all—people speak of the past as being behind them, while the future lies ahead. They visualize it that way, too. "Forgetting what is behind and straining toward what is ahead, I press on," says Paul. To imagine the future or the past as a "place" is already to engage in analogy. Are there "places" in time, as there are in space? To say so is to assert that time is *like* space. *The past is a foreign country: they do things differently there.* The future, too. If time is a fourth dimension, that is because it is *like* the first three: visualizable as a line; measurable in extent. Still, in other ways time is *unlike* space. The fourth dimension differs from the other three. They do things differently there.

It seems natural to sense time as a spacelike thing. Accidents of language encourage that. We have only so many words; *before* and *after* have to do double duty as prepositions of space as well as time. "Time is a phantasm of motion," said Thomas Hobbes in 1655. To count time, to compute time, "we make use of some motion or other, as of the sun, of a clock, of the sand in an hourglass." Newton considered time to be absolutely different from space—after all, space *remains always immovable,* whereas *time flows equably without regard to anything external, and by another name is called duration*—but his mathematics created an inevitable analogy between time and space. You could plot them as axes on a graph. By the nineteenth century German philosophers in particular were groping toward some amalgam of time and space. Arthur Schopenhauer wrote in 1813, "In mere Time, all things *follow one another,* and in mere Space all things are *side by side;* it is accordingly only by the combination of Time and Space that the representation of coexistence arises." Time as a dimension begins to emerge from the mists. Mathematicians could see it. Technology helped in another way. Time became vivid, concrete, and spatial to anyone who saw the railroad smashing across distances on a coordinated schedule—coordinated by the electric telegraph, which was pinning time to the mat. "It may seem strange to 'fuse' time and space," explained the *Dublin Review,* but look—here is a "quite ordinary" space-time diagram:

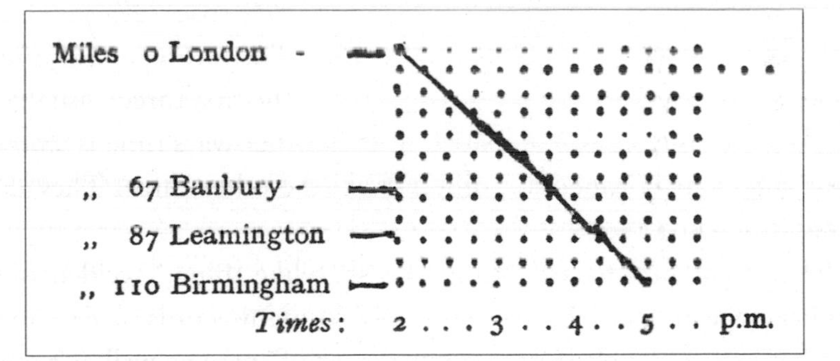

So Wells's Time Traveller can speak with conviction: "Scientific people know very well that Time is only a kind of Space. Here is a popular scientific diagram, a weather record. This line I trace with my finger shows the movement of the barometer. . . . Surely the mercury did not trace this line in any of the dimensions of Space . . . but certainly it traced such a line, and that line, therefore, we must conclude was along the Time-Dimension."

In the new century everything felt new; physicists and philosophers gazed upon Time, so often capitalized, with new eyes. Twenty-five years after *The Time Machine* the "new realist" philosopher Samuel Alexander put it this way:

> If I were asked to name the most characteristic feature of the thought of the last twenty-five years I should answer: the discovery of Time. I do not mean that we have waited till to-day to become familiar with Time. I mean that we have only just begun in our speculation to take Time seriously and to realize that in some way or other Time is an essential ingredient in the constitution of things.

What is time? Time machines may help us understand.

WELLS WAS NOT READING Schopenhauer, and philosophical introspection was not his style. His ideas about time were informed by Lyell and Darwin, who read the buried strata that frame the ages of the earth and the ages of life. He studied zoology and geology as a scholarship student at the Normal School of Science and Royal School of Mines, and these subjects encouraged him to view the world's history as if from a great height—its lost epochs, a panorama unfolding, "the small-scale horse-foot, hand-industry civilizations that culminated in the seventeenth and eighteenth centuries, by the change of pace and scale due to mechanical invention." Geological time, so vastly extended, disrupted the earlier sense of historical time, in which the world was plausibly considered to be six thousand years old. The scales were so different; human history was dwarfed.

"O earth, what changes has thou seen!" wrote Tennyson. "The hills are shadows, and they flow/From form to form, and nothing stands." Lately, too, there was a science called archeology—grave robbers and treasure hunters in the service of knowledge. Archeologists, digging, were exposing buried history. At Nineveh, at Pompeii, at Troy, vaults were opened; past civilizations appeared, frozen in stone, but lifelike. Archeological digs exposed diagrams ready-made, with time a visible dimension.

Less obviously, people could see layers of time all around. Travelers riding in steam-driven railroad trains looked out their windows onto a landscape where oxen plowed the fields as they had done in medieval times, horses still hauled and harrowed, yet telegraph wires split the sky. This caused a new kind of confusion or dissociation. Call it temporal dissonance.

Above all, modern time was irreversible, inexorable, and unrepeatable. Progress marched onward—a good thing, if you were a technological optimist. Cyclical time, crosswinds of time, eternal return, the wheel of life: these were romantic notions now, for poets and nostalgic philosophers.

The Normal School, later renamed the Royal College of Science, was a lucky place for H. G., the youngest son of a shopkeeper and a former housemaid. As a teenager he had spent three unhappy years serving as a draper's apprentice. Now, in the college's new five-story

elevator-equipped building, he studied elementary biology with ("under the shadow of") Thomas H. Huxley, the famous Darwinian—a mighty intellectual liberator, Wells thought, bravely battling the priests and know-nothings, establishing the facts of evolution from painstakingly assembled fossil evidence and embryological material, filling up the "great jig-saw puzzle," the confirmation of the tree of life. It was the most educational year of his life: "a grammar of form and a criticism of fact." He had less use for the course in physics, of which he later remembered little but his own ineptitude in trying to contrive a barometer from some bits of brass and wood and glass tubing.

After finishing at the Normal School, he supported himself with some schoolteaching before "collapsing" (his word) into literary journalism. Here he found an outlet for the kind of high-flown scientific speculation he had enjoyed in the Debating Society. One essay for the *Fortnightly Review,* "The Rediscovery of the Unique," grandly assessed "the series of dissolving views that we call the march of human thought." His next, titled "The Universe Rigid," was declared incomprehensible by the review's formidable editor, Frank Harris, who summoned the twenty-four-year-old author to his office and tossed the manuscript into the trash bin. The Universe Rigid was a construct of four dimensions—like a block. It does not change over time, because time is already built in.

The four-dimensional frame led as if by iron necessity to the Universe Rigid. If you believed in the laws of physics in those days—and the students of the Normal School in the nation of Newton most assuredly did—then apparently the future must be a strict consequence of the past. Wells proposed to design a "Universal Diagram" by which all phenomena would be logically deduced.

> One began with a uniformly distributed ether in the infinite space of those days and then displaced a particle. If there was a Universe rigid, and hitherto uniform, the character of the consequent world would depend entirely, I argued along strictly materialist lines, upon the velocity of this initial displacement.

And then? Chaos!

> The disturbance would spread outward with ever increasing complication.

Edgar Allan Poe, similarly inspired by scientific speculation, wrote in 1845, "As no thought can perish, so no act is without infinite result." In a story called "The Power of Words," published in the *Broadway Journal,* he invents some angels who explain:

> We moved our hands, for example, when we were dwellers on the earth, and, in so doing, we gave vibration to the atmosphere which engirdled it. This vibration was indefinitely extended, till it gave impulse to every particle of the earth's air, which thenceforth, *and for ever,* was actuated by the one movement of the hand. This fact the mathematicians of our globe well knew.

The actual mathematician Poe had in mind was the arch-Newtonian Pierre-Simon, Marquis de Laplace, for whom the past and the future were nothing more or less than physical states, rigidly connected by the inexorable mechanics of the laws of physics. The present state of the universe (he wrote in 1814) is "the effect of its past and the cause of its future." Here is the Universe Rigid:

> Given for one instant an intelligence which could comprehend all the forces by which nature is animated and the respective positions of the beings which compose it, if moreover this intelligence were vast enough to submit these data to analysis, it would embrace in the same formula both the movements of the largest bodies in the universe and those of the lightest atom; to it nothing would be uncertain, and the future as the past would be present to its eyes.

Some people already believed in such an intelligence; they called it "God." To Him nothing would be uncertain or unseen. Doubt is for us mortals. The future, as the past, would be present to His eyes. (Or would it? Perhaps God would be content to see creation unfold. Heaven's virtues might include patience.)

This one sentence by Laplace has more enduring life than the rest of his work combined. It pops up again and again in the philosophizing of the next two centuries. Whenever anyone starts talking about fate or free will or determinism, there is the marquis again. Jorge Luis Borges mentions his "fantasies": "that the present state of the universe is, in theory, reducible to a formula, from which Someone could deduce the entire future and the entire past."

The Time Traveller invents "an omniscient observer":

> To an omniscient observer there would be no forgotten past—no piece of time as it were that had dropped out of existence—and no blank future of things yet to be revealed. Perceiving all the present, an omniscient observer would likewise perceive all the past and all the inevitable future at the same time. Indeed, present and past and future would be without meaning to such an observer: he would always perceive exactly the same thing. He would see, as it were, a Rigid Universe filling space and time—a Universe in which things were always the same.*

"If 'past' meant anything," he concludes, "it would mean looking in a certain direction; while 'future' meant looking the opposite way."

The Universe Rigid is a prison. Only the Time Traveller can call himself free.

* This passage appears in an early serialized version in the *New Review* (volume 12, page 100) but not in the final book.

TWO

Fin de Siècle

Your body moves always in the present, the dividing line between the past and the future. But your mind is more free. It can think, and is in the present. It can remember, and at once is in the past. It can imagine, and at once is in the future, in its own choice of all the possible futures. Your mind can travel through time!

—Eric Frank Russell (1941)

CAN YOU, citizen of the twenty-first century, recall when you first heard of time travel? I doubt it. Time travel is in the pop songs, the TV commercials, the wallpaper. From morning to night, children's cartoons and adult fantasies invent and reinvent time machines, gates, doorways, and windows, not to mention time ships and special closets, DeLoreans, and police boxes. Animated cartoons have been time traveling since 1925: in "Felix the Cat Trifles with Time," Father Time agrees to send the unhappy Felix back to a faraway time inhabited by cavemen and dinosaurs. In a 1944 Looney Tunes episode, Elmer Fudd dreams his way into the future—"when you hear the sound of the gong it will be exactly 2000 AD"—where a newspaper headline reveals, "Smellevision Replaces Television." By 1960 *Rocky and His Friends* was sending the dog Mr. Peabody and his adopted boy, Sherman, through the WABAC Machine to straighten out William Tell and Calamity Jane, and the next year Donald Duck made his first trip into prehistory, to invent the wheel. "Wayback Machine" became a thing, so a sitcom character says, "Dave, don't mess with a man with a Wayback Machine—I can make it so you were never born."

Children learn about "time whirlwinds" and "time-travel stones." Homer Simpson accidentally turns a toaster into a time machine. No explanation is necessary. We've outgrown the need for professors expounding on the fourth dimension. What's not to understand?

China's official State Administration of Radio, Film, and Television issued a warning and denunciation of time travel in 2011, concerned that such stories interfere with history—"casually make up myths, have monstrous and weird plots, use absurd tactics, and even promote feudalism, superstition, fatalism, and reincarnation." Indeed. Global culture has absorbed the tropes of time travel. In *The Onion,* a photograph of a man with a futuristic-looking e-cigarette occasions an article about a time-traveling "soldier of fortune with off-world military training." People can work out his whole story just by looking at him. "Judging by his cool, calm demeanor and the fact that he was inhaling what looked like e-fumes from some kind of shiny black mecha-cigarette, I'm just going to assume this guy has journeyed here from hundreds of years in the future to apprehend a dangerous digi-convict of some kind," says an onlooker. "Imagine his knowledge of future events. He could probably share information about so many astounding secrets if we dared ask." Others reckon his sunglasses hide advanced ocular cybernetics and that he's traversing the space-time continuum armed with a pulse rifle or particle cannon. "Further sources speculated, with growing alarm, that the man's very presence in the bar might somehow cause an irreversible temporal paradox of some kind."

Nor does time travel belong solely to popular culture. The time-travel meme is pervasive. Neuroscientists investigate "mental time travel," more solemnly known as "chronesthesia." Scholars can hardly broach the metaphysics of change and causality without discussing time travel and its paradoxes. Time travel forces its way into philosophy and infects modern physics.

Have we spent the last century developing a lurid pipe dream? Have we lost touch with the simple truth about time? Or is it the other way around: perhaps the blinders have come off and we are finally evolving, as a species, an ability to understand the past and the future for what they are. We have learned a great deal about time, and only some of it from science.

HOW STRANGE, then, to realize that time travel, the concept, is barely a century old. The term first occurs in English in 1914*—a back-

* Per the *Oxford English Dictionary.* One precursor, though: in 1866, an English travel writer concluding a railway journey through Transylvania mused in the *Cornhill Magazine,* "This charm of traveling would become perfect if we could travel in time as well as in space— . . . take a fortnight in the fifteenth century, or, still more pleasant, a leap into the twenty-first. It is possible to accomplish this object more or less in imagination."

formation from Wells's "Time Traveller." Somehow humanity got by for thousands of years without asking, What if I could travel into the future? What would the world be like? What if I could travel into the past—could I change history? The questions didn't arise.

By now *The Time Machine* is one of those books you feel you must have read at some point, whether or not you actually did. You may have seen the 1960 movie, starring the matinee-handsome Rod Taylor as the Time Traveller (he needed a name, so they called him George) and featuring a machine that didn't remind anyone of a bicycle. Bosley Crowther in the *New York Times* called this time machine "an antique version of the flying saucer." To me it looks like a rococo sort of sledge, with a plush red chair. Apparently I'm not the only one. "Everyone knows what a time machine looks like," writes the physicist Sean Carroll: "something like a steampunk sled with a red velvet chair, flashing lights, and a giant spinning wheel on the back." The movie also features the Time Traveller's erstwhile companion, Weena, played by Yvette Mimieux as a languid peroxide blonde of the year 802701.

George asks Weena whether her people think much about the past. "There is no past," she informs him, with no discernible conviction. Do they wonder about the future? "There is no future." She lives in the now, all right. Everyone has forgotten about fire, too, but luckily George brought some matches. "I'm only a tinkering mechanic," he says modestly, but he'd like to fill her in on a few things.

Motion picture technology, by the way, was just coming over the horizon when Wells wrote his fantasy, and he took note. (The bicycle was not the only modern machine from which he drew inspiration.) In 1879 the photographic stop-motion pioneer Eadweard Muybridge invented what he called a zoopraxiscope for projecting successive images to give the illusion of movement. They made visible an aspect of time never before seen. Thomas Edison followed with his kinetoscope and met in France with Étienne-Jules Marey, who was already creating *la chronophotographie,* followed soon after by Louis and Auguste Lumière and their *cinématographe.* By 1894, London was entertaining crowds at its first kinetoscope parlor in Oxford Street; Paris had one, too. So when the Time Traveller begins his voyage it looks like this:

> I pressed the lever over to its extreme position. The night came like the turning out of a lamp, and in another moment came to-morrow. The laboratory grew faint and hazy, then fainter and ever fainter. To-morrow night came black, then day again, night again, day again, faster and faster still. An eddying murmur filled my ears, and a strange, dumb confusedness descended on my mind. . . . The twinkling succession of darkness and light was excessively painful to the eye. Then, in the intermittent darknesses, I saw the moon spinning swiftly through her quarters from new to full, and had a faint glimpse of the circling stars. Presently, as I went on, still gaining velocity, the palpitation of night and day merged into one continuous greyness.

One way or another, the inventions of H. G. Wells color every time-travel story that followed. When you write about time travel, you either pay homage to *The Time Machine* or dodge its shadow. William Gibson, who would reinvent time travel yet again in the twenty-first century, was a boy when he encountered Wells's story in a fifteen-cent

Classics Illustrated comic book; by the time he saw the movie he felt he already owned it, "part of a personal and growing collection of alternate universes."

> I had imagined this, for my own purposes, as geared in some achingly complex spheres-within-spheres way that I could never envision in operation. . . . I suspected, without admitting it to myself, that time travel might be a magic on the order of being able to kiss one's own elbow (which had seemed, initially, to be quite theoretically possible).

In his seventy-seventh year Wells tried to recall how it came to him. He couldn't. He needed a time machine for his own consciousness. He put it almost that way himself. His brain was stuck in its epoch. The instrument doing the recollecting was also the instrument to be recalled. "I have been trying, for a day or so, to reconstruct the state of my brain as it was about 1878 or 9. . . . I find it impossible to disentangle. . . . The old ideas and impressions were made over in accordance with new material, they were used to make up the new equipment." Yet if ever a story was kicking to be born, it was *The Time Machine.*

It flowed from his pen in fits and starts over a period of years, beginning in 1888 as a fantasy called "The Chronic Argonauts," serialized in three installments in the *Science Schools Journal,* a periodical Wells started himself at the Normal School. He rewrote it and threw it away at least twice. A few dramatic early bits survive: "Conceive me, the Time Traveller, the discoverer of Futurity"—futurity!—"clinging senseless to his Time Machine, choking with sobs & with the tears streaming down his face, full of a terrible fear that he would never see humanity again."

In 1894 he revived "that old corpse," as it already seemed, for a series of seven anonymous pieces in the *National Observer* and then produced a nearly final version, at last called *The Time Machine,* for serial publication in the *New Review.* The hero was called Dr. Moses Nebogipfel, Ph.D., F.R.S., N.W.R., PAID—"a small-bodied, sallow-faced little man . . . aquiline nose, thin lips, high cheek-ridges, and pointed chin . . . his extreme leanness . . . large eager-looking grey eyes . . . phenomenally wide and high forehead." Nebogipfel turned into the Philosophical Inventor and then into the Time Traveller. But he did not so much evolve as fade. He lost his honorary initials and even his name; he shed all the lively word painting and became nondescript, a gray spectre.

Naturally it seemed to Bertie that he was the one striving: learning his craft, shredding his drafts, rethinking and rewriting late into the night by the light of a paraffin lamp. He struggled, certainly. But let's say instead that the story was in charge. The time for time travel had come. Donald Barthelme suggests we see the writer as "the work's way of getting itself written, a sort of lightning rod for an accumulation of atmospheric disturbances, a St. Sebastian absorbing in his tattered breast the arrows of the Zeitgeist." That may sound like a mystical metaphor or a bit of false modesty, but a lot of writers talk that way and they seem to mean it. Ann Beattie says Barthelme is giving away an inside secret:

> Writers don't talk to nonwriters about being hit by lightning, being conduits, being vulnerable. Sometimes they talk that way to each

> other, though. The work's way of getting itself written. I think that's an amazing concept that not only gives words (the work) a mind and a body but gives them the power to stalk a person (the writer). Stories do that.

Stories are like parasites finding a host. In other words, memes. Arrows of the Zeitgeist.

"Literature is revelation," said Wells. "Modern literature is indecorous revelation."

THE OBJECT OF Wells's interest, bordering on obsession, was the future—that shadowy, inaccessible place. "So with a kind of madness growing upon me, I flung myself into futurity," says the Time Traveller. Most people, Wells wrote—"the predominant type, the type of the majority of living people"—never think about the future. Or, if they do, they regard it "as a sort of blank non-existence upon which the advancing present will presently write events." (The moving finger writes; and having writ, moves on.) The more modern sort of person—"the creative, organizing, or masterful type"—sees the future as our very reason for being: "Things have been, says the legal mind, and so we are here. The creative mind says we are here because things have yet to be." Wells, of course, hoped to personify that creative, forward-looking type. He had more and more company.

In bygone times, people had no more than the barest glimmerings of visiting either the future or the past. It seldom occurred to anyone. It wasn't in the repertoire. Even travel through space was rare, by modern standards, and slow, before the railroads came.

If we stretch, we can find arguable cases of precocious time travel.

Time travel *avant la lettre*. In the Mahabharata, the Hindu epic, Kakudmi ascends to the heavens to meet Brahma and finds upon his return that epochs have passed and everyone he knew is dead. A similar fate befalls an ancient Japanese fisherman, Urashima Tarō—an inadvertent leap into the future by journeying far from home. Likewise Rip Van Winkle could be said to have accomplished time travel by sleeping. There was also time travel by dreaming, time travel by hallucinogen, or time travel by mesmerism. The nineteenth-century literature includes one instance of time travel by message in a bottle: by none other than Poe, who described "an odd-looking MS." that he found "corked up in a jug" floating in an imaginary sea and bearing the dateline "ON BOARD BALLOON 'SKYLARK' April 1, 2848."

Aficionados have scoured the attics and basements of literary history for other examples—time-travelish precursors. In 1733, an Irish clergyman, Samuel Madden, published a book called *Memoirs of the Twentieth Century*: an anti-Catholic diatribe in the form of letters from British officials living two hundred years hence. The twentieth century as imagined by Madden resembles his own time in every respect except that Jesuits have taken over the world. The book was unreadable even in 1733. Madden destroyed almost all of the thousand copies himself. A handful remain. By contrast, a utopian vision titled *L'an deux mille quatre cent quarante: rêve s'il en fût jamais* (The Year 2440: A Dream If There Ever Was One) became a sensational bestseller in prerevolutionary France. It was a utopian fantasy published in 1771 by Louis-Sébastien Mercier, heavily influenced by the philosopher of the hour, Rousseau. (The historian Robert Darnton puts Mercier in the category of *Rousseaus du ruisseau,* or "gutter Rousseaus.") His narrator dreams that he has awakened from a long sleep to find he has acquired wrinkles and a large nose. He is seven hundred years old and about to discover the Paris of the future. What's new? Fashion has changed—people wear loose clothes, comfortable shoes, and odd caps. Societal mores have changed, too. Prisons and taxes have been abolished. Society abhors prostitutes and monks. Equality and reason prevail. Above all, as Darnton points out, a "community of citizens" has eradicated despotism. "In imagining the future," he says, "the reader could also see what the present would look like when it had become the past." But Mercier, who believed that the earth was a flat plain under an orbiting sun, was not looking toward the year 2440 so much as the year 1789. When the Revolution came, he declared himself to have been its prophet.

Another vision of the future, also utopian in its way, appeared in 1892: a book titled *Golf in the Year 2000; or, What Are We Coming To,* by a Scottish golfer named J. McCullough (given name lost in the mists). When the story begins, its narrator, having endured a day of bad golf and hot whiskies, falls into a trance. He awakens wearing a heavy beard. A man solemnly tells him the date. "'It is' (and he referred to a pocket almanac as he spoke) 'the twenty-fifth of March, 2000.'" Yes, the year 2000 has advanced to pocket almanacs. Also electric lights. In some respects, though, the golfer from 1892 discovers that the world evolved while he slept. In the year 2000 women dress like men and do all the work, while men are freed to play golf every day.

Time travel by hibernation—the long sleep—worked for Washington Irving in "Rip Van Winkle," and for Woody Allen in his 1973 remake, *Sleeper.* Woody Allen's hero is Rip Van Winkle with a modern set of neuroses: "I haven't seen my analyst in two hundred years. He was a strict Freudian. If I'd been going all this time, I'd probably almost be cured by now." Is it a dream or a nightmare, if you open your eyes to find your contemporaries all dead?

Wells himself dispensed with the machinery in a 1910 novel, *The Sleeper Awakes,* which was also the first time-travel fantasy to discover the benefits of compound interest. Anyway, sleeping into the future is what we do every night. For Marcel Proust, five years younger than Wells and two hundred miles away, no place heightened the awareness of time more than the bedchamber. The sleeper frees himself from time, floats outside of time, and drifts between insight and perplexity:

> A sleeping man holds in a circle around him the sequence of the hours, the order of the years and worlds. He consults them instinctively as he wakes and reads in a second the point on the earth he occupies, the time that has elapsed before his waking; but their ranks can be mixed up, broken. . . . In the first minute of his waking, he will no longer know what time it is, he will think he has only just gone to bed. . . . Then the confusion among the disordered worlds will be complete, the magic armchair will send him traveling at top speed through time and space.

Traveling, that is, metaphorically. In the end, the sleeper rubs his eyes and returns to the present.

Machines improved upon magic armchairs. By the last years of the nineteenth century, novel technology was impressing itself upon the culture. New industries stirred curiosity about the past as well as the future. So Mark Twain created his own version of time travel in 1889, when he transported a Connecticut Yankee into the medieval past. Twain didn't worry about scientific rationalization, but he did frame the story with some highfalutin verbiage: "You know about transmigration of souls; do you know about transposition of epochs—and bodies?" For *A Connecticut Yankee in King Arthur's Court* the means of time travel is a bang on the head: Hank Morgan, the Yankee, gets beaned with a crowbar and awakens in a verdant field. Before him sits an armor-clad fellow on a horse, wearing (the horse, that is) festive red and green silk trappings like a bed quilt. Just how far the Connecticut Yankee has traveled he discovers in this classic exchange:

> "Bridgeport?" said I, pointing.
>
> "Camelot," said he.

Hank is a factory engineer. This is important. He is a can-do guy and a technophile, up-to-date on the latest inventions: blasting powder and speaking-tubes, the telegraph and the telephone. So was the author. Samuel Clemens installed Alexander Graham Bell's telephone in his home in 1876, the year it was patented, and two years before

A WORD OF EXPLANATION. 21

me out with a crusher alongside the head that made everything crack, and seemed to spring every joint in my skull and make it overlap its neighbor. Then the world went out in darkness, and I didn't feel anything more, and didn't know anything at all—at least for a while.

When I came to again, I was sitting under an oak tree, on the grass, with a whole beautiful and broad country land-that he acquired an extraordinary writing machine, the Remington typewriter. "I was the first person in the world to apply the typemachine to literature," he boasted. The nineteenth century saw wonders.

The steam age and the machine age were in full swing, the railroad was shrinking the globe, the electric light turning night into never-ending day, the electric telegraph annihilating time and space (so the newspapers said). This was the true subject of Twain's *Yankee:* the contrast of modern technology with the agrarian life that came before. The mismatch is both comic and tragic. Foreknowledge of astronomy makes the Yankee a wizard. (The nominal wizard, Merlin, is exposed as a humbug.) Mirrors, soap, and matches inspire awe. "Unsuspected by this dark land," Hank says, "I had the civilization of the nineteenth century booming under its very nose!" The invention that seals his triumph is gunpowder.

What magic might the twentieth century bring? How medieval might we seem to the proud citizens of that future? A century earlier, the year 1800 had passed with no special fanfare; no one imagined how different the year 1900 might be.* Time awareness in general was dim, by our sophisticated standards. There is no record of a "centennial" celebration of anything until 1876. (The *Daily News,* London, reported, "America has been of late very much centennialised—that is the word in use now since the great celebration of this year. Centennials have been got up all over the States.") The expression "turn of the

* Of course the century was turning only per the Christian calendar, and even so, in 1800, the consensus was barely firm. France, still in the throes of its revolution, was running on a new calendar of its own, *le calendrier républicain français,* so it was the year 9. Or 10. This Republican year had a neat 360 days, organized into months with new names, from *vendémiaire* to *fructidor.* Napoleon dispensed with that shortly after being crowned emperor on 11 *frimaire,* year 13.

century" didn't exist until the twentieth. Now, finally, the Future was becoming an object of interest.

The New York industrialist John Jacob Astor IV published a "romance of the future" six years before the turn of the century, titled *A Journey in Other Worlds.* In it he forecast myriad technological developments for the year 2000. Electricity, he predicted, would replace animal power for the movement of all vehicles. Bicycles would be fitted with powerful batteries. Enormous high-speed electric "phaetons" would roam the globe, attaining speeds as great as thirty-five to forty miles an hour on country roads and "over forty" on city streets. To support these carriages, pavement would be made of half-inch sheets of steel laid over asphalt ("though this might be slippery for horses' feet, it never seriously affects our wheels"). Photography would be wonderfully advanced, no longer limited to black and white: "There is now no difficulty in reproducing exactly the colours of the object taken."

In Astor's year 2000, telephone wires girdle the earth, kept underground to avoid interference, and telephones can show the face of the speaker. Rainmaking has become "an absolute science": clouds are manufactured by means of explosions in the upper atmosphere. People can soar through space to visit the planets Jupiter and Saturn, thanks to a newly discovered antigravitational force called "apergy"—"whose existence the ancients suspected, but of which they knew so little." Does that sound exciting? It all seemed "terribly monotonous" to the reviewer for the *New York Times:* "It is a romance of the future, and it is as dull as a romance of the Middle Ages." It was Astor's fate, too, to go down with the *Titanic.*

As a vision of an idealized world, a sort of utopia, Astor's book owed a debt to Edward Bellamy's *Looking Backward,* the American bestseller of 1887, likewise set in the year 2000. (Time travel by sleep again:

our hero enters a trance of 113 years.) Bellamy expressed the frustration of not being able to know the future. In his story "The Blindman's World" he imagined that we earthlings are alone among the universe's intelligent creatures in lacking "the faculty of foresight," as if we had eyes only at the backs of our heads. "Your ignorance of the time of your death impresses us as one of the saddest features of your condition," says a mysterious visitor. *Looking Backward* inspired a wave of utopias, to be followed by dystopias, and these are so invariably futuristic that we sometimes forget that the original *Utopia,* by Thomas More, was not set in the future at all. Utopia was just a faraway island.

No one bothered with the future in 1516. It was indistinguishable from the present. However, sailors were discovering remote places and strange peoples, so remote *places* served well for speculative authors spinning fantasies. Lemuel Gulliver does not voyage in time. It is enough for him to visit "Laputa, Balnibarbi, Luggnagg, Glubbdubdrib, and Japan." William Shakespeare, whose imagination seemed limitless, who traveled freely to magical isles and enchanted forests, did not—could not—imagine different *times.* The past and present are all the same to Shakespeare: mechanical clocks strike the hour in Caesar's Rome, and Cleopatra plays billiards. He would have been amazed by the theatrical time travel that Tom Stoppard creates in *Arcadia* and *Indian Ink*: placing together on stage stories that unfold in different eras, decades apart.

"Something needs to be said about this," Stoppard writes in a stage direction in *Arcadia*. "The action of the play shuttles back and forth between the early nineteenth century and the present day, always in this same room." Props move about—books, flowers, a tea mug, an oil lamp—as if crossing the centuries through an invisible portal. By the end of the play they have gathered on a table: *the geometrical solids, the computer, decanter, glasses, tea mug, Hannah's research books, Septimus's books, the two portfolios, Thomasina's candlestick, the oil lamp, the dahlia, the Sunday papers* . . . On Stoppard's stage these objects are the time travelers.

We have achieved a temporal sentience that our ancestors lacked. It was long in coming. The year 1900 brought a blaze of self-consciousness about times and dates. The twentieth century was rising like a new sun. "No century has ever issued from the womb of time whose advent has aroused the high expectation, the universal hope, as that which the midnight litanies and the secular festivals but eight days hence will usher in," wrote the editorialist of the *Philadelphia Press.* The Hearst-owned *New York Morning Journal* declared itself "The Twentieth Century Newspaper" and organized an electrical publicity stunt: "The Journal Asks All Citizens of New York to Illuminate Their Homes Monday Midnight as a Welcome to the Twentieth Century." New York festooned City Hall with two thousand lightbulbs in red, white, and blue, and the president of the city council addressed a throng: "Tonight when the clock strikes twelve the present century will have come to an end. We look back upon it as a cycle of time within which the achievements in science and in civilization are not less than marvelous." In London the *Fortnightly Review* invited its now famous futurist, the thirty-three-year-old H. G. Wells, to write a series of prophetic essays: "Anticipations of the Reaction of Mechanical and Scientific Progress upon Human Life and Thought." In Paris they were already calling it fin de siècle, emphasis on *fin:* decadence and ennui were all the rage. But when the time came, the French, too, looked forward.

An English writer could not hope to have an international literary reputation until he had been published in France, and Wells did not have to wait long. *The Time Machine* was translated by Henry Davray, who recognized an heir to the visionary Jules Verne, and the venerable *Mercure de France* printed it in 1898 with a title that lost something in translation: *La machine à explorer le temps.** Naturally the avant-garde loved the idea of time travel: Avant! Alfred Jarry, a symbolist playwright and prankster—also an enthusiastic bicyclist—using the pseudonym "Dr. Faustroll," immediately produced a mock-serious construction manual, *"Commentaire pour servir à la construction pratique de la machine à explorer le temps."* Jarry's time machine is a bicycle with an ebony frame and three "gyrostats" with rapidly rotating flywheels, chain drives, and ratchet boxes. A lever with an ivory handle controls the speed. Mumbo-jumbo ensues. "It is worth noting that the Machine has two Pasts: the past anterior to our own present, what we might call the real past, and the past created by the Machine when it returns to our Present and which is in effect the reversibility of the Future." Time is the fourth dimension, of course.† Jarry later said he admired Wells's "admirable sang-froid" in managing to make *his* mumbo-jumbo so scientific.

The fin de siècle was at hand. Preparing for Year 1900 festivities in Lyon, Armand Gervais, a toy manufacturer who liked novelties and automata, commissioned a set of fifty color engravings from a

* Evidently it was not easy to translate. *Current Literature* magazine in New York reported in 1899, "The 'Mercure de France' is about to begin publishing a translation of Mr. Wells' *Time Machine.* The translator finds the title difficult to put into French. *'Le Chronomoteur,' 'Le Chrono Mobile,' 'Quarante Siècles à l'heure,'* and *'La Machine à Explorer le Temps'* are some of the suggestions. . . ."

† Jarry explains: "The Present is non-existent, a tiny fraction of a phenomenon, smaller than an atom. The physical size of an atom is known to be 1.5×10^{-8} centimeters in diameter. No one has yet measured the fraction of a solar second that is equal to the Present."

freelance artist named Jean-Marc Côté. These images conjure a world of marvels that might exist *en l'an 2000:* people sporting in their tiny personal aircraft, warring in dirigibles, playing underwater croquet at the bottom of the sea. Perhaps the best is the schoolroom, where children in knee breeches sit with hands clasped at wooden desks while their teacher feeds books into a grinding machine, powered by a hand crank. Evidently the books are pulverized into a residue of pure information, which is then conveyed by wires up the wall and across the ceiling and down into headsets that cover the pupils' ears.

These prescient images have a story of their own. They never saw the light of their own time. A few sets were run off on the press in the basement of the Gervais factory in 1899, when Gervais died. The factory was shuttered, and the contents of that basement remained hidden for the next twenty-five years. A Parisian antiques dealer stumbled upon the Gervais inventory in the twenties and bought the lot, including a single proof set of Côté's cards in pristine condition. He had them for fifty years, finally selling them in 1978 to Christopher Hyde, a Canadian writer who came across his shop on rue de l'Ancienne-Comédie. Hyde, in turn, showed them to Isaac Asimov, a Russian-born scientist and science-fiction writer, the author or editor of, by then, 343 books. Asimov made the *En l'an 2000* cards into his 344th: *Futuredays.* He saw something remarkable in them—something genuinely new in the annals of prophecy.

Prophecy is old. The business of "telling" the future has existed through all recorded history. Foretelling and soothsaying, augury and divination, are among the most venerable of professions, if not always the most trusted. Ancient China had 易經, *I Ching,* the *Book of Changes;* sibyls and oracles plied their trade in Greece; aeromancers and palmists and scryers saw the future in clouds, hands, and crystals, respectively. "That grim old Roman Cato the Censor said it well: 'I wonder how one augur can keep from laughing when he passes another,'" wrote Asimov.

But *the future,* as divined by the diviners, remained a personal matter. Fortune-tellers cast their hexagrams and turned their tarot cards to see the futures of individuals: sickness and health, happiness and misery, tall dark strangers. As for the world at large—that did not change. Through most of history, the world people imagined their children living in was the world they inherited from their parents. One generation was like the next. No one asked the oracle to forecast the character of daily life in years to come.

"Suppose we dismiss fortune-telling," says Asimov. "Suppose we also dismiss divinely inspired apocalyptic forecasts. What, then, is left?"

Futurism. As redefined by Asimov himself. H. G. Wells talked about "futurity" at the turn of the century, and then the word *futurism* was hijacked by a group of Italian artists and protofascists. Filippo

Tommaso Marinetti published his "Futurist Manifesto" in the winter of 1909 in *La Gazzetta dell'Emilia* and *Le Figaro,* declaring himself and his friends to be free at last—free of the past.

> An immense pride was buoying us up, because we felt ourselves alone at that hour, alone, awake, and on our feet, like proud beacons or forward sentries against an army of hostile stars. . . . *"Andiamo,"* I said. *"Andiamo, amici!"* . . . And like young lions we ran after death, [etc.]

The manifesto included eleven numbered items. Number one: "We intend to sing the love of danger . . ." Number four was about fast cars: "We affirm that the world's magnificence has been enriched by a new beauty: the beauty of speed. A racing car whose hood is adorned with great pipes, like serpents of explosive breath." The *futuristi* created just one of the many twentieth-century movements that proudly defined themselves as avant-garde—eyes fixed forward, escaping the past, striding into the future.

When Asimov used the word, he meant something more basic: a sense of the future as a notional place, different, and perhaps profoundly different, from what has come before. Through most of history, people could not see the future that way. Religions had no particular thought for the future; they looked toward rebirth, or eternity—a new life after death, an existence outside of time. Then, finally, humanity crossed a threshold of awareness. People began to sense that there *was* something new under the sun. Asimov explains:

> Before we can have futurism, we must first recognize the existence of the future in a state that is significantly different from the present

> and the past. It may seem to us that the potential existence of such a future is self-evident, but that was most definitely not so until comparatively recent times.

And when did that happen? It began in earnest with the Gutenberg printing press, saving our cultural memory in something visible, tangible, and shareable. It reached critical velocity with the Industrial Revolution and the rise of the machine—looms and mills and furnaces, coal and iron and steam—creating, along with so much else, a sudden nostalgia for the apparently vanishing agrarian way of life. Poets led the way. *"Hear the voice of the Bard!"* William Blake implored, *"Who Present, Past, and Future sees."* Some people liked progress more than did Mr. Dark Satanic Mills, but either way, before futurism could be born, people had to *believe* in progress. Technological change had not always seemed like a one-way street. Now it did. The children of the Industrial Revolution witnessed vast transformations within their lifetimes. To the past there was no return.

Surrounded by advancing machinery, Blake blamed, more than anyone else, Isaac Newton—the blinkered rationalist imposing his new order*—but Newton himself had not believed in progress. He studied a great deal of history, mostly biblical, and if anything he supposed that his own era represented a fall from grace, a tattered remnant of past glories. When he invented vast swaths of new mathematics, he thought he was rediscovering secrets known to the ancients and later forgotten. His idea of absolute time did not subvert his belief in eternal Christian time. Historians studying our modern notion of

* "May God us keep/From Single vision & Newton's sleep!"

progress have observed that it began to develop in the eighteenth century, along with our modern notion of history itself. We take our sense of history for granted—our sense of "historical time." The historian Dorothy Ross defines it as "the doctrine that all historical phenomena can be understood historically, that all events in historical time can be explained by prior events in historical time." (She calls this "a late and complex achievement of the modern West.") It seems so obvious now: we build upon the past.

So, as the Renaissance receded, a few writers began trying to imagine the future. Besides Madden with his *Memoirs of the Twentieth Century* and Mercier with his dream of the year 2440, others attempted imaginative fiction about societies to come, which can, in hindsight, be called "futuristic," though that word did not register in English until 1915. They were all defying Aristotle, who wrote, "Nobody can *narrate* what has not yet happened. If there is narration at all, it will be of past events, the recollection of which is to help the hearers to make better plans for the future."

The first true futurist in Asimov's sense of the word was Jules Verne. In the 1860s, as railroad trains chugged across the countryside and sailing ships gave way to steam, he imagined vessels traveling under the sea, across the skies, to the center of the earth, and to the moon. We would say he was a man ahead of his time—he had an awareness, a sensibility, suited to a later era. Edgar Allan Poe was ahead of his time. The Victorian mathematician Charles Babbage and his protégé Ada Lovelace, forerunners of modern computing, were ahead of their time. Jules Verne was so far ahead of his time that he could never even find a publisher for his most futuristic book, *Paris au XXe siècle,* a dystopia featuring gas-powered cars, "boulevards lit as brightly as by the sun," and machine warfare. The manuscript, handwritten in a yellow notebook, turned up in 1989, when a locksmith cracked open a long-sealed family safe.

The next great futurist was Wells himself.

We are all futurists now.

THREE

Philosophers and Pulps

"Time travel?! You expect me to believe such nonsense?"

"Yes, it is a difficult concept, isn't it."

—Douglas Adams ("The Pirate Planet," *Doctor Who,* 1978)

TIME TRAVEL AS DESCRIBED by Wells and his many heirs is everywhere now, but it does not exist. It cannot. In saying so, it occurs to me that I'm Filby.

"But the thing's a mere paradox," says the Editor.

"It's against reason," says Filby.

Critics of the 1890s took the same view. Wells knew they would. When his book was finally published in the spring of 1895—*The Time Machine: An Invention,* sold in New York by Henry Holt (75¢) and in London by William Heinemann (2/6)—reviewers admired it for a good tale: a "fantastic story"; "shocker of no ordinary kind"; "*tour de force* of ghastly imaginings"; "distinctly above the average of such fanciful works"; and "worth reading if you like to read impossible yarns" (that last from the *New York Times*). They noted the apparent influence of the Dark Romantics, Edgar Allan Poe and Nathaniel Hawthorne. One sniffed, "We have some difficulty in discerning the exact utility of such excursions into futurity."

Only a few paid Wells the compliment of analyzing his fantastic notion logically. They found it illogical. "There is no getting into the Future, except by waiting," wrote Israel Zangwill in the *Pall Mall Magazine,* wagging a stern finger. "You can only sit down and see it come by." Zangwill, himself an occasional novelist and humorist, soon also to be a famous Zionist, thought he understood time quite well. He admonished the author:

> In verity there is no Time Traveller, Mr. Wells, save Old Father Time himself. Instead of being a Fourth Dimension of Space, Time is perpetually travelling through Space, repeating itself in vibrations farther and farther from the original point of incidence; a vocal panorama moving through the universe across the infinities, a succession of sounds and visions that, having once been, can never pass away . . .

(Zangwill had clearly been reading Poe: vibrations indefinitely extending through the atmosphere—"no thought can perish"—and this sentence, too, sails indefinitely onward.)

> . . . but only on and on from point to point, permanently enregistered in the sum of things, preserved from annihilation by the endlessness of Space, and ever visible and audible to eye or ear that should travel in a parallel movement.

Despite his objections Zangwill couldn't help but admire Wells's "brilliant little romance." He shrewdly noticed that *The Arabian Nights* had employed a sort of time-machine precursor: a magic carpet that traverses space. Meanwhile, even in 1895, Zangwill seemed to understand certain peculiar implications of time travel—the paradoxes, we will soon say—better than Wells himself.

The Time Machine looks one way: forward. Ostensibly Wells's time machine could travel to the past with a reverse throw of the lever, but the Time Traveller had no interest in going there. And a good thing too, says Zangwill. Just think what difficulties would be entailed. Our past had no Time Traveller barging through. A past that included a Time Traveller would be a different past, a new past. None of this was easy to put into words:

> Had he travelled backwards, he would have reproduced a Past which, in so far as his own appearance in it with his newly invented machine was concerned, would have been *ex hypothesi* unveracious.

Then there's the problem of meeting oneself. Zangwill becomes the first to notice this, and he will not be the last:

> Had he recurred to his own earlier life, he would have had to exist in two forms simultaneously, of varying ages—a feat which even Sir Boyle Roche would have found difficult.

(His readers would recognize Roche as the Irish politician who said, "Mr. Speaker, it is impossible I could have been in two places at once, unless I were a bird.")*

* Sir Boyle is also remembered for this: "Why should we put ourselves out of our way to do anything for posterity, for what has posterity ever done for us?"—a joke that reads differently now that we have time travel. Posterity does plenty for us: sends us assassins and bounty hunters on covert missions to change the course of history, for example.

The book reviewers came and went, and before long philosophers got into the game. When they first took notice of time travel it was with a certain embarrassment, like a symphony conductor unable to look away from an organ grinder. "A frivolous example drawn from contemporary fiction," said Professor Walter Pitkin at Columbia University, writing in the *Journal of Philosophy* in 1914. Something was stirring in science—the realm in which time was a measurable and absolute quantity known familiarly as *t*—and philosophers were uneasy. In the first years of the new century, when they turned to the subject of time, they had one thinker above all to contend with: the young Frenchman Henri Bergson. In the United States, William James, who might otherwise have been resting on his laurels as the "father of psychology," found new vitality in Bergson. "Reading his works is what has made me bold," James said in 1909. "If I had not read Bergson, I should probably still be blackening endless pages of paper privately, in the hope of making ends meet that were never meant to meet." (He added, "I have to confess that Bergson's originality is so profuse that many of his ideas baffle me entirely.")

Bergson asks us to remember how artificial is the notion of space as an empty homogenous medium—the absolute space announced by Newton. It is a creation of human intellect, he notes: "We may as well say that men have a special faculty of perceiving or conceiving a space without quality." Scientists may find this abstract empty space to be useful for calculating, but let's not make the mistake of confusing it with *reality.* Even more so with time. When we measure time with mechanical clocks, when we draw diagrams where time is an axis on a graph, we may fall into the trap of imagining time to be merely another version of space. To Bergson, time *t,* the time of the physicists, sliced into hours, minutes, and seconds, turned philosophy into a prison. He rejected the immutable, the absolute, the eternal. He embraced flux, process, becoming. For Bergson, the philosophical analysis of time could not be divorced from our human experience of it: the overlapping of mental states, the segue from one to the next that we experience as duration—*la durée.*

He held time apart from space rather than commingling the two: "Time and space begin to interweave only when both become fictional." He saw time, not space, as the essence of consciousness; duration, the heterogeneous tide of moments, as the key to freedom. Philosophers were about to follow physics down a new path, and Bergson would be left behind, but for now he was hugely popular. Crowds thronged his lectures at the Collège de France, Proust attended his wedding, and James called him a magician. "Dive back into the flux then," cried James. "Turn your face toward sensation, that flesh-bound thing which rationalism has always loaded with abuse." Here he parted company with physics.

> What really *exists* is not things made but things in the making. Once made, they are dead. . . . Philosophy should seek this kind of living understanding of the movement of reality, not follow science in vainly patching together fragments of its dead results.

Pitkin seems to have felt that he needed to rescue the poor scientists from Bergson's onslaught. Described by *Time* magazine in a brief moment of fame as "a man of many ideas, some of them large," he was a founding member of a short-lived movement that called itself new realism. In his 1914 essay he declared that he liked some of Bergson's "conclusions" but despised his "entire method," particularly the rejection of scientific process in favor of psychological introspection. Pitkin proposed to clear up the space-time conundrum by means of logical proof. He would embrace the physicists' t and t' and t'' and yet he would prove once and for all that time is different from space. To wit: we can move hither and yon in space, but not in time. Or rather, we do move in time, but not freely: "a thing moves in time only by moving with all other things." And how would he prove this? In a most unexpected way:

> To make the proof as simple as possible, I shall present it in the form of a sober criticism of one of the wildest flights of literary fancy which that specialist in wild flights, H. G. Wells, has indulged in. I refer, of course, to his amusing skit, *The Time Machine.*

It was the first but not the last time that Mr. Wells's amusing skit would impose itself upon the attention of this august journal.

"You cannot leap back into the thirteenth century, nor can a man of that period hop into our own," wrote Pitkin. "Mr. Wells would have us imagine a man at rest in the space dimensions, but moving with respect to the time of that space field. Very well! Let us do our noblest to play the game. What do we find? Something very disconcerting indeed. Something which, I fear, will make time-touring very unpopular among sedate people."

> The traveler flies, not through abstract time (like the "pure space" of the geometer). *He flies through real time. But real time is history: and history is the course of physical events. It is the sequence of activities, physical, physiological, political, and otherwise.*

Do we really want to go down this road? Must we look for errors of logic in a piece of fantastic fiction?

Yes, we must. The practitioners of time travel, even in "pulp" magazines, were soon to work out rules and justifications that would make a Talmudist proud. What is allowed, what is possible, what is plausible—the rules evolved and varied, but logic must be honored. We may as well begin with Professor Pitkin, man of many ideas, some of them large, in the *Journal of Philosophy*.

His argument would not seem very sophisticated to a typical teenage sci-fi aficionado circa 1970. In fairness, he recognizes that common human intuition about the world often fails to comprehend the strangeness of reality. Science keeps surprising us. Which way is up, for example? "It was held impossible 'by the very nature of things,'" he notes, "that the earth should be a sphere, with people on the other side walking, heads downward." (He might have added that Aristotle's common sense revealed three and no more than three spatial dimensions: "the line has magnitude in one way, the plane in two ways, and the solid in three ways, and beyond these there is no other magnitude because the three are all.") Could it be, he asks, that time travel merely strikes us as impossible "because of certain prejudices we entertain or certain facts and tricks of which we are still hopelessly ignorant?" Let us be open-minded. "[The] answer, whatever that may be, carries immeasurable consequences for metaphysics."

So Pitkin applies the tools of logic. These are his chief points:

- As the time machine rushes through the years, everything ages rapidly, so the man in the machine should, too. "Nations rise and fall, tempests leap up, destroy, and subside, houses are built with toil and burned in the frenzy of sudden war, and so on." As for the tourist, his clothes are unruffled and he scarcely ages a day. "How is this possible? If he has passed through a hundred thousand generations, why isn't he a hundred thousand generations old?" Here is an obvious contradiction: "the first contradiction in the whole proceeding."

• Time goes at a certain rate, and this rate must be the same for everyone, everywhere. "Two objects or systems" cannot have "different rates of displacement or change in time"—obviously. Pitkin scarcely knew what devilishness Albert Einstein was conjuring in Berlin.

• Traveling through time must obey rules of arithmetic, just like traveling through space. Do the math: "To traverse a million years in a few days is exactly like traveling a thousand miles in one inch." A thousand miles does not equal one inch; ergo, a million years cannot equal a few days. "Now is not this a pure self-contradiction, on a par with the proposition that you or I can go from New York to Pekin without moving farther than our own front door?"

• The time traveler would surely bump into things. Example: Let's say he leaves his workshop for a future date, January 1, 1920. While he's gone, his abandoned wife sells the house. It is torn down. Bricks are heaped where the workshop stood. "But where, oh where, is the traveler? If he remains in the same place, he is surely beneath the ton of bricks and so is his precious machine. . . . This, we aver, is most uncomfortable for the tourist. He is fairly interpenetrated with bricks."

• Looked at from an astronomical point of view, celestial motion must be considered as well. "The traveler who moves only in time and not at all in space would suddenly find himself strangling in the empty ether, while the earth went hurtling away from beneath him."

Impossible, concludes the philosopher. No one can travel into the future or the past on Mr. Wells's time machine. We must find other ways of dealing with past and future, every day of our lives.

WE NEED NOT DEFEND Mr. Wells, because he never meant to promulgate a new theory of physics. He didn't believe in time travel. The time machine was the handwaving—the pixie dust that helps the willing reader suspend disbelief and get through the story. (See *handwavium, n.*) It was sheer coincidence that the Time Traveller's mumbo-jumbo tracked so well with a revolutionary view of spacetime that emerged in physics a decade later. Except, of course, that it was no coincidence at all.

Wells worked hard to make the handwaving plausible. This first technology of time travel ended up being fairly robust. In fact, he anticipated Pitkin's semiscientific objections and some others as well. For example, it is the Medical Man who says that space differs from time in that we move freely through one but not the other.

"Are you so sure we can move freely in space?" the Time Traveller retorts. "Right and left we can go, backward and forward freely enough. . . . But how about up and down? Gravitation limits us there." That was more true, of course, in the nineteenth century than in the twenty-first. Now we're used to whizzing about in all three of our dimensions, but *space travel* (as we might call it) used to be more constrained. Railroads and bicycles were new. So were elevators and balloons. "Before the balloon," says the Time Traveller, "save for spasmodic jumping and the inequalities of the surface, man had no freedom of vertical movement." What the balloon does for the third dimension, the time machine might do for the fourth.

Our hero presents his miniature prototype time machine as an amalgam of science and magic: "You will notice that it looks singularly askew, and that there is an odd twinkling appearance about this bar, as though it was in some way unreal." A turn of the tiny lever sends the gizmo into the void with a puff of wind. Now Wells anticipates the next objection from the realists. If the time machine has gone back to the past, why had they not seen it en route (as it were) when they were in the room last Thursday? And if into the future, why is it not still visible, passing through each successive moment? The explanation comes in ersatz psychological jargon. "It's presentation below the threshold," says the Time Traveller, nodding to the Psychologist. "You know, diluted presentation." The same reason you can't see the spokes of a spinning bicycle wheel or a bullet whizzing through the air. ("Of course," the Psychologist replies. "I should have thought of it.")

Wells likewise foresaw the objection of the Philosopher that the traveler risked crashing through piles of bricks and other unexpected alterations in the landscape. "So long as I travelled at a high velocity through time, this scarcely mattered; I was, so to speak, attenuated—was slipping like a vapour through the interstices of intervening substances!" Simple, when you put it that way. Halting in the wrong place, however, would still be dangerous. And exciting.

> To come to a stop involved the jamming of myself, molecule by molecule, into whatever lay in my way; meant bringing my atoms into such intimate contact with those of the obstacle that a profound chemical reaction—possibly a far-reaching explosion—would result,

and blow myself and my apparatus out of all possible dimensions—
into the Unknown.

Wells laid down the rules, and from then on all the world's time travelers would have to obey. Or if not obey, at least explain. Jack Finney put it this way in a time-travel story in the *Saturday Evening Post,* 1962: "There's a danger a man might appear in a time and place already occupied. . . . He'd be all mixed in with the other molecules, which would be unpleasant and confining." Explosions are ever popular. Philip K. Dick in 1974: ". . . the hazard in re-entry of being out of phase spatially and colliding right down to the molecular level with two tangent objects. . . . You know, 'No two objects can occupy the same space at the same time.'" At last, the perfect corollary of "Nobody can be in two places at once."

Wells never did justify treating the earth as a fixed point of the cosmos. Nor for that matter did he worry about where the Time Machine gets the energy to power its voyages. Here, too, he established a tradition. Even a bicycle needs someone to pedal, but time machines have unlimited free fuel, by the universe's grace.

WE'VE HAD A CENTURY to think about it, and we still need to remind ourselves every so often that time travel is not real. It's an impossibility, just as William Gibson suspected—a magic on the order of kissing one's own elbow. But when I say that to a certain well-known theoretical physicist, he gives me a pitying look. Time travel is no problem, he says. At least if you want to travel to the future.

Oh, well, sure—you mean we're all traveling forward in time anyway?

No, says the physicist, not just that. Time travel is easy! Einstein showed us how to do it. All we have to do is approach a black hole and accelerate to near the speed of light. Then, welcome to the future.

His point is that acceleration and gravitation both slow the clocks, relativistically, so you could age a year or two on a spaceship and return home a century hence to marry your great-grandniece (as Tom Bartlett does in Robert Heinlein's 1956 novel *Time for the Stars*). This is proven. GPS satellites have to compensate for relativistic effects in their very exact calculations. It's hardly time travel, though. It is time dilation (per Einstein, *Zeitdilatation*). It's an antiaging device.* And it's a one-way street. There's no going back to the past. Unless you can find a wormhole.

"Wormhole" is John Archibald Wheeler's word for a shortcut through the warped fabric of spacetime—a "handle" of multiply connected space. Every few years someone makes headlines by hailing the possibility of time travel through a wormhole—a traversable wormhole, or maybe even a "macroscopic ultrastatic spherically-symmetric long-throated traversable wormhole." I believe that these physicists have been unwittingly conditioned by a century of science fiction. They've read the same stories, grown up in the same culture as the rest of us. Time travel is in their bones.

We have arrived at a moment of cultural history when the doubters and naysayers are the real practitioners of time travel, the science-fiction writers themselves. "Totally impossible on theoretical grounds,"

* When the American astronaut Scott Kelly returned to Earth in March 2016 after nearly a year of high-speed orbit, he was reckoned to be 8.6 milliseconds younger, relative to his groundling twin brother, Mark. (Then again, Mark had lived through only 340 days while Scott experienced 10,944 sunrises and sunsets.)

declared Isaac Asimov in 1986. He didn't even bother to hedge his bets.

> It can't and won't be done. (If you're one of those romantics who thinks nothing is impossible, I won't argue the case, but I trust you won't decide to hold your breath until such a machine is built.)

Kingsley Amis, assessing the literary culture of science fiction in 1960, felt he was stating the obvious when he said, "Time travel, for instance, is inconceivable." Thus practitioners of the genre resort to some version of Wells's hand-waving explanation—"an apparatus of pseudo-logic"—or, as time goes on, simply trust their readers to suspend disbelief. And so it's the science-fiction writers who remain willing to treat the future as open, while all around them physicists and philosophers surrender to determinism. "One is grateful that we have a form of writing which is interested in the future," said Amis, "which is ready to treat as variables what are usually taken to be constants."

As for Wells himself, he continued to disappoint his believers.* "The reader got a fine confused sense of immense and different

* J. B. Priestley, who loved Wells and credited him with inspiring his Time Plays, said, "Although he was never rude about it he deplored the way in which I was bothering my head about Time in the thirties. He was like a man who, having wrongly given up playing an instrument for which he had a flair, then refused to listen to anybody else playing it." Another disappointed admirer, W. M. S. Russell, echoed Priestley's complaint at a centennial symposium in 1995: "More than a century after his wonderful achievement, let us be remembering, not the disillusioned elder, but the young creator of *The Time Machine*."

things," he said in 1938. "The effect of reality is easily produced. One jerks in one or two little unexpected gadgets or so, and the trick is done. It is a trick." (He was just back in London after a seven-city American lecture tour titled "Organization of the World Brain," and he felt a need to deny special futuristic powers. "It is not a bit of good pretending I am a prophet. I have no crystal into which I gaze, and no clairvoyance.")

LET'S LOOK one more time at how the trick was done:

> *. . . the dance of the shadows, how we all followed him, puzzled but incredulous, and how there in the laboratory we beheld a larger edition of the little mechanism which we had seen vanish from before our eyes. Parts were of nickel, parts of ivory, parts had certainly been filed or sawn out of rock crystal. The thing was generally complete, but the twisted crystalline bars lay unfinished upon the bench beside some sheets of drawings, and I took one up for a better look at it. Quartz it seemed to be.*
>
> *"Look here," said the Medical Man, "are you perfectly serious? Or is this a trick . . . ?"*

For Wells's first readers, technology had a special persuasive power. This vague machine put a claim on the readers' belief in a way that magic never could. Magic might include clouts on the head, as in *Connecticut Yankee,* as well as the talismanic act of turning back the hands on a clock. The cartoon "Felix the Cat Trifles with Time" employs both devices: Old Father Time unwinds his clock past "Year of 1" and "Stone Age" *and* whacks poor Felix with a club.

Before that, in 1881, a newspaperman, Edward Page Mitchell, published "The Clock That Went Backward" anonymously in the *New York Sun.* Old Aunt Gertrude, spectral in her white nightgown and white nightcap, has a mysterious bond with her eight-foot-tall Dutch clock. It seems defunct—until one night, when she winds it up in the flickering light of a candle, the hands begin to turn backward, and she falls dead. This becomes the occasion for a philosophical disquisition by one Professor Van Stopp:

> Well, and why should not a clock go backward? Why should not Time itself turn and retrace its course? . . . Viewed from the Absolute, the sequence by which future follows present and present follows past is purely arbitrary. Yesterday, today, tomorrow; there is no reason in the nature of things why the order should not be tomorrow, today, yesterday.

If the future is different from the past, what if we reverse the mirror or rewind the clock? Can destiny carry us toward our beginnings? Can effect influence cause?

The device of the backward-running clock reappeared in a 1919 story, "The Runaway Skyscraper," by the pseudonymous Murray Leinster. "The whole thing started when the clock on the Metropolitan Tower began to run backward" is its opening sentence. The tower trembles, the office workers hear ominous creaking and groaning, the sky darkens, night falls, the telephones produce only static, and all too soon the sun rises again, at high speed, and in the west.

"Great bombs and little cannon-balls," shouts Arthur, a young engineer who has been worrying about his debts. "It looks awfully queer," agrees Estelle, his twenty-one-year-old secretary, who has been worrying that she will become "an old maid." The landscape transforms at a rapid pace, wristwatches are seen spinning backward, and finally Arthur puts two and two together. "I don't know how to explain it," he explains. "Have you ever read anything by Wells? *The Time Machine,* for instance?"

Estelle shakes her head no. "I don't know how I'm going to say it so you'll understand," explains Arthur manfully, "but time is just as much a dimension as length and breadth." The building has "settled back in the Fourth Dimension," he decides. "We're going back in time."

These stories were multiplying. Another way to do the trick: bring in the devil. "A tall, flashy, rather Mephistophelean man whom I had seen from time to time in the domino-room" makes his appearance in Max Beerbohm's "Enoch Soames," published in the *Century* illustrated magazine in 1916. Enoch Soames is a "dim" man, stooping and shambling, an unsuccessful striver in 1890s literary London. He is, like some other writers, concerned about how posterity will remember him. "A hundred years hence!" he cries. "Think of it! If I could come back to life then—just for a few hours . . ."

That is the devil's cue, of course. He offers a bargain—the Faustian kind, updated.

"Parfaitement," he says Frenchily. "Time—an illusion. Past and future—they are as ever present as the present, or at any rate only what you call 'just round the corner.' I switch you on to any date. I project you—pouf!"

The devil is au courant: like everyone else, he has been reading *The Time Machine*. "But it is one thing to write about an impossible machine," he says. "It is a quite other thing to be a supernatural power." The devil says pouf and poor Enoch gets his wish. Transported to 1997, he materializes in the Reading Room of the British Museum and heads straight for the S volumes of the card catalogue. (How better to gauge one's literary reputation?) There he learns his fate: "Enoch Soames," he reads, was an imaginary character in a 1916 story by a mordant writer and caricaturist named Max Beerbohm.

IN THE TWENTIES the future seemed to be arriving daily. News had never traveled so fast, and there had never been so much of it, with the advent of wireless transmission, and by 1927 Wells himself had already had enough. The technology of communications had reached maturity, he felt, with wireless telegraphy, the wireless telephone, "and all the broadcasting business." Radio had begun as a glorious dream—the finest fruits of the culture, the wisest thoughts and best music, transmitted into homes across the land. "Chaliapin and Melba would sing to us, President Coolidge and Mr. Baldwin would talk to us simply, earnestly, directly; the most august in the world would wish us good evening and pass a friendly word; should a fire or shipwreck happen, we were to get the roar of the flames and the cries for help." A. A. Milne would tell stories to children and Albert Einstein would bring science to the masses. "All sporting results before we went to bed would be included, the weather forecast, advice about our gardens, the treatment of influenza, and the exact time."

Yet for Wells the dream had turned sour. Asked by the *New York Times* to assess the state of radio for its readers, he ranted bitterly, disillusioned as a child finding lumps of coal in the Christmas stocking. "Instead of first-rate came tenth-rate music, played by the Little Winkle-Beach Pier Band," he wrote. Instead of the wisest voices, "Uncle Bray and Aunt Twaddle." Even the static irritated him. "Across it all dear old Mother Nature cast her net of 'atmospherics' with a humor all her own." He did enjoy hearing a bit of dance music after a long day—"but dance music only goes on for a small part of the evening, and at any moment it may give way to Doctor Flatulent being thoughtful and kindly in a non-sectarian way."

His assessment was so harsh that the *Times* editors were clearly taken aback. They emphasized that Wells could speak of radio broadcasting only "as He Encounters It Abroad." Wells was not only disappointed in the present state of radio. His crystal ball showed him that the whole enterprise was doomed to fade away. "The future of broadcasting is like the future of crossword puzzles and Oxford trousers, a very trivial future indeed." Why would anyone listen to music by radio when they could have gramophone records? Radio news vanishes like smoke: "Broadcasting shouts out its information once and cannot be recalled." For serious thought, he said, nothing can replace books.

His Majesty's Government had created a "salaried official body to preside over broadcasting programs," Wells noted—the new British Broadcasting Company. "In the end that admirable committee may find itself arranging schemes of entertainment for a phantom army of expiring listeners." If any audience remained at all, it would comprise "the blind, lonely and suffering people"—or "probably very sedentary persons living in badly lighted houses or otherwise unable to read, who have never realized the possibilities of the gramophone and the pianola and who have no capacity for thought or conversation." The BBC's first experimental television broadcasts were just five years away.

Others could play the futurism game, though. David Sarnoff of RCA retorted by calling Wells a snob; the inventor Lee de Forest told him he needed a better radio—and perhaps the most unusual rebuttal came from the publisher of *Radio News* and manager of station WRNY, an émigré from Luxembourg named Hugo Gernsback. After arriving in New York at the age of nineteen, Gernsback had started the Electro Importing Company, a mail-order business selling radio parts to eager hobbyists in 1905, with tantalizing advertisements in *Scientific American* and elsewhere. Within three years he was printing his own magazine, *Modern Electrics*. By the twenties he was well known to legions of radio amateurs. "I refuse to believe in such a drab and dreary demise of radio," he wrote in a letter to the *Times*. "What surprises me most is that the prophetic Mr. Wells has not looked into the near future when every radio set will be equipped with its television attachment—a device, by the way, now being perfected by one of his own countrymen." (This was not the only thing that surprised him most. "What surprises me most in Mr. Wells's remarks," he said in the same letter, "is that he evidently hankers to listen constantly to the great, when a simple mathematical calculation would show that this would not be possible. There are not enough great people in the world.")

Gernsback was an extraordinary person: a self-made inventor, an entrepreneur, and what people of a later time would term a bullshit artist. Around town he wore expensively tailored suits, used a monocle to examine the wine lists of expensive restaurants, and ran nimbly from creditors. When one of his magazines failed, two more would rise up. *Radio News* was not destined to be the most influential of his magazines, nor was *Sexology,* the "Illustrated Magazine of Sex Science." The Gernsback creation that mattered most to future history was a so-called pulp magazine—named for its cheap wood-pulp paper—sold for twenty-five cents an issue, called *Amazing Stories.* Its rough pages made room for a variety of advertisements: "450 Miles on a Gallon of Gas," free sample from Whirlwind Mfg. Co. of Milwaukee; "Correct Your Nose, shapes flesh and cartilage while you sleep, 30 Day Trial Offer, Free Booklet"; and "New Scientific Wonder: X-Ray Curio, Boys, Big Fun, You apparently see thru Clothes, Wood, Stone, any object. See Bones in Flesh, price 10¢." He found a ready market for what he was selling. He lectured to New York audiences about the marvels of the future and broadcast his lectures live on WRNY, and the *New York Times* reported them breathlessly. "Science will find ways to transmit tons of coal by radio, facilitate foot traffic by electrically propelled roller skates, save electric current by cold light and grow and harvest crops electrically, according to a forecast of the next fifty years made by Hugo Gernsback," the paper declared in 1926. Weather control would be complete, and city skyscrapers would all have flat tops for landing airplanes.

> Huge high frequency electric current structures, placed on top of our largest buildings, will either dispel threatening rain, or, if necessary, produce rain as needed, during the hot spells or during the night. . . . We may soon expect fantastic towers piercing the sky and giving off weird purple glows at night when energized. . . . Fifty years hence you will be able to see what is going on in your favorite broadcast station, and you will meet your favorite singer face to face. You will watch the Dempsey of fifty years hence battle with his Tunney, whether you are on board an airship or away in the wilds of Africa, or such wilds as still exist.

By the end of his life he had eighty patents to his name. He anticipated radar as early as 1911.

Then again, he arranged what he claimed was the first-ever "entirely successful" test of hypnotism by radio: the hypnotist, Joseph Dunninger, who also served as head of the department of magic for Gernsback's *Science and Invention* magazine, put a subject named Leslie B. Duncan into a trance from a distance of ten miles. The *Times* reported that, too: "Duncan's body was then placed over two chairs, forming a human bridge, and Joseph H. Kraus, field editor of *Science and Invention,* was able to sit on the improvised bridge."

All this came under the rubric of fact. For fiction, he had *Amazing Stories.*

Beginning in April 1926, *Amazing Stories* was the first periodical solely devoted to a genre that did not, until this moment, have a name. In Paris in 1902, Alfred Jarry wrote an admiring essay about the "scientific novel" or "hypothetical novel"—the novel that asks, "What if . . . ?" The hypothetical novel might later prove futuristic,

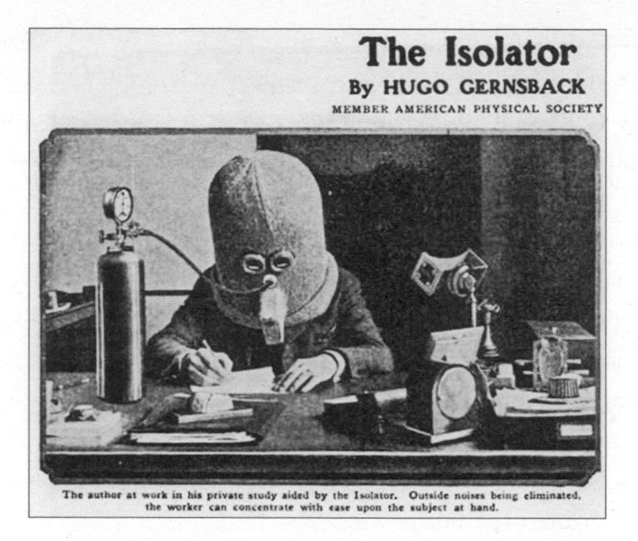

The author at work in his private study aided by the Isolator. Outside noises being eliminated, the worker can concentrate with ease upon the subject at hand.

he suggested, depending on the future. Maurice Renard, a practitioner himself, declared this a whole new genre, which he called "the scientific-marvelous novel" (*le roman merveilleux scientifique*). "I say a new genre," he wrote in *Le Spectateur;* after all, *genre* was a French word. "Until Wells," he added, "one might well have doubted it."

Gernsback dubbed it "scientifiction." "By 'scientifiction,'" he wrote in the first issue, "I mean the Jules Verne, H. G. Wells, and Edgar Allan Poe type of story—a charming romance intermingled with scientific fact and prophetic vision." He had published quite a few of these before, even in *Radio News,* and had written a serial novel of his own, *Ralph 124C*[*] *41+: A Romance of the Year 2660* (self-published in his *Modern Electrics* magazine and described by Martin Gardner

* Spoken aloud: "One to foresee . . ."

much later as "surely the worst SF novel ever written").* It took just a few more years for "scientifiction" to become "science fiction." Gernsback lost control of *Amazing Stories* in one of his bankruptcies, but the magazine continued for almost eighty years and helped define the genre. "Extravagant Fiction Today—Cold Fact Tomorrow" was the magazine's motto.

"Let it be understood," Gernsback wrote in a short treatise for would-be writers, "that a science fiction story must be an exposition of a scientific theme and it must be also a story. . . . It must be reasonable and logical and must be based upon known scientific principles."† In the first issues of *Amazing Stories* he reprinted Verne, Wells, and Poe, along with Murray Leinster's "Runaway Skyscraper." In the second year he reprinted the entire *Time Machine.* He didn't bother paying for these reprints. He offered writers twenty-five dollars for original stories, but they often had trouble collecting. As part of his tireless promotion of the genre, Gernsback founded a fan organization, the Science Fiction League, with chapters in three countries.

So the idea of science fiction as a genre, distinct from literary fiction and presumably inferior, was born here, in trashy magazines barely distinguishable from the funnies or pornography. Yet so was a

* Kingsley Amis also took the time to read this book. "*Ralph 124C 41+* concerns the technological marvels invented or demonstrated by the ridiculously resourceful eponymous hero. . . . After some trouble with a pair of rival suitors, one human, the other Martian, Ralph restores a dead girl to life by a complicated deep-freeze and blood-transfusion technique. Other wonders include the hypnobioscope . . . and three-dimensional color television, a term which Gernsback is credited, if that is the word, with having invented."

† He also proposed a few "don'ts," including, "Don't make your professor, if you have one, talk like a military policeman or an Eighth Avenue 'cop.' Don't put cheap jokes in his mouth. Read semi-technical magazines and reports of speeches to get the flavor of academic phraseology."

cultural form, a way of thinking, that soon could not be dismissed as trash. "I can just suggest," wrote Kingsley Amis when not much time had passed, "that while in 1930 you were quite likely to be a crank or a hack if you wrote science fiction, by 1940 you could be a normal young man with a career to start, you were a member of the first generation who had grown up with the medium already in existence." In the pages of the pulps, the theory and praxis of time travel began to take shape. Besides the stories themselves, there were letters from probing readers and notes from the editors. Paradoxes were discovered and, with some difficulty, put into words.

"How about this *Time Machine*?" wrote "T.J.D." in July 1927. Consider some other possibilities. What if our inventor journeys back to his schoolboy days? "His watch ticks forward although the clock on the laboratory wall goes backward." What if he encounters his younger self? "Should he go up and shake hands with this 'alter ego'? Will there be two physically distinct but characteristically identical persons? . . . Boy! Page Einstein!"

Two years later Gernsback had a new scientifiction magazine, this time called *Science Wonder Stories,* sister publication to *Air Wonder Stories,* and the December 1929 issue featured on its cover a story of time travel called "The Time Oscillator."* It involved, yet again, some odd machinery with crystals and dials and some professorial discourse on the fourth dimension. ("As I have before explained, time is only a relative term. It means literally nothing.") This time the trav-

* An editor's note explained: "Stories of traveling in time are always exceedingly interesting reading, mainly for the reason that the feat has not yet been accomplished; though no one can say that it cannot be done in the future, when we have reached a much higher plane of scientific achievement. Traveling in time, either forward or backward, may well become a possibility."

elers head off into the distant past—which prompted a special editor's note from Gernsback. "Can a time traveler," he asked, "going back in time—whether ten years or ten million years—partake in the life of that time and mingle in with its people; or must he remain suspended in his own time-dimension, a spectator who merely looks on but is powerless to do more?" A paradox loomed; Gernsback could see it plainly, and he put it into words:

> Suppose I can travel back into time, let me say 200 years; and I visit the homestead of my great great great grandfather. . . . I am thus enabled to shoot him, while he is still a young man and as yet unmarried. From this it will be noted that I could have prevented my own birth; because the line of propagation would have ceased right there.

Henceforth this would be known as the grandfather paradox. It turns out that one person's objection is another's story idea. Gernsback invited comments from readers by mail and received quite a few, over a period of years. A boy in San Francisco suggested yet another paradox, "the last knock on time traveling": What if a man were to travel into the past and marry his mother? Could he be his own father?

Page Einstein indeed.

FOUR

Ancient Light

"Time is a mental concept," said Pringle. "They looked for time everywhere else before they located it in the human mind. They thought it was a fourth dimension. You remember Einstein."

—Clifford D. Simak (1951)

BEFORE WE HAVE clocks we experience time as fluid, mercurial, and inconstant. Pre-Newtonians did not assume that time was a universal, trustworthy, absolute affair. Time was well known to be relative—to use that word in its psychological sense, not to be confused with the newer sense that came into being circa 1905. *Time travels in divers paces with divers persons.** Clocks reified time and then Newton made time . . . let's say, official. He made it an essential part of science: time *t*, a factor to be plugged into equations. Newton regarded time as part of the "sensorium of God." His view is handed down to us as if engraved on tablets of stone:

> Absolute, true, and mathematical time, in and of itself and of its own nature, without reference to anything external, flows uniformly . . .

* Rosalind adds: "I'll tell you who time ambles withal, who time trots withal, who time gallops withal, and who he stands still withal."

The cosmic clock ticks invisibly and inexorably, everywhere the same. Absolute time is God's time. This was Newton's credo. He had no evidence for it, and his clocks were rubbish compared to ours.

> It may be, that there is no such thing as an equable motion, whereby time may be accurately measured. All motions may be accelerated and retarded, but the flowing of absolute time is not liable to any change.

Besides religious conviction, Newton was motivated by mathematical necessity: he needed absolute time, as he needed absolute space, in order to define his terms and express his laws. Motion is defined as the change in place over time; acceleration is the change in velocity over time. With a backdrop of absolute, true, and mathematical time, he could build an entire cosmology, a System of the World. This was an abstraction; a convenience; a framework for calculating. But for Newton it was also a statement about the world. You may believe it, or not.*

Albert Einstein believed it. Up to a point.

He believed in an edifice of laws and computation that had grown from a bare stone church into a grand ornate cathedral, supported by colonnades and flying buttresses, layered with carving and tracery—work still in progress, with hidden crypts and ruined chapels. In this edifice time t played an indispensable part. No one could grasp the

* The philosopher and physicist Ernst Mach, a forebear of relativity, objected to absolute time in 1883: "It is utterly beyond our power to measure the changes of things by time. . . . Time is an abstraction at which we arrive by means of the changes of things." Einstein quoted that approvingly when he wrote Mach's obituary in 1916, but he himself could not go so far in expunging the convenient abstraction. Time remained an essential property of his universe.

whole structure, but Einstein understood more than most and had encountered a problem. There was an internal contradiction. The great achievement of the last century's physics was James Clerk Maxwell's unification of electricity, magnetism, and light—the achievement that was so visibly wiring the whole world. Electric currents, magnetic fields, radio waves, and light waves were one and the same. Maxwell's equations made it possible to calculate the speed of light, for the first time. But they were not meshing perfectly with the laws of mechanics. Those light waves, for example—so clearly waves, according to the mathematics, but waves *in what*? Sound needs air or water or other substance to carry the vibrations. Light waves likewise implied an unseen medium, the so-called ether—"luminiferous," or light bearing. Naturally experimentalists were trying to detect this ether, with no success. Albert Michelson and Edward Morley came up with a clever experiment in 1887 to measure the difference between the speed of light in the direction of the earth's motion and the speed of light at right angles to it. They couldn't find any difference at all. Was the ether necessary? Or was it possible to think purely of an electrodynamics of moving bodies, through empty space?

We know now that the speed of light in empty space is constant, 299,792,458 meters per second. No rocket ship can overtake a flash of light or reduce that number in the slightest. Einstein struggled ("psychic tension"; "all sorts of nervous conflicts") to make sense of that: to discard the luminiferous ether, to accept the speed of light as absolute. Something else had to give. On a fine bright day in Bern (as he told the story later), he talked it over with his friend Michele Besso. "Next day I came back to him again and said to him, without even saying hello, 'Thank you. I've completely solved the problem.' *An analysis of the concept of time was my solution.*" If light speed is absolute, then time itself cannot be. We must abandon our faith in perfect simultaneity: the assumption that two events can be said to happen at the same time. Multiple observers experience their own present moments. "Time cannot be absolutely defined," said Einstein—it can be defined, but not *absolutely*—"and there is an inseparable relation between time and signal velocity."

The signal carries information. Suppose six sprinters line up at the start line for the hundred-meter run, with their hands and one knee touching the ground and their feet in the starting blocks, awaiting the sound of the gun. The signal velocity in this case will be about a few hundred meters per second, the speed of sound through air. That's slow nowadays, so Olympic events have scrapped starting pistols in favor of signals wired (at light speed) into loudspeakers. To think about simultaneity more carefully, it becomes necessary also to consider the signal velocity of light traveling to the eyes of the runners, the judges, and the spectators. In the end, there is *no* one instant, no "point in time," that can be the same for everyone.

Suppose lightning strikes a railway embankment (trains are more usual than horses in these stories) at two different points, distant from each other. Can you—a physicist, with the most excellent modern equipment—establish whether the two flashes were simultaneous? You cannot. It turns out that a physicist riding the train will disagree with a physicist standing at the station. Every observer owns a reference frame, and each reference frame has its own clock. There is no one cosmic clock, no clock of God or Newton.

The revelation is that we can share no *now*—no universal present moment. But was that altogether a surprise? Before Einstein was born, John Henry Newman, poet and priest, wrote that "time is not a common property;/But what is long is short, and swift is slow/And near is distant, as received and grasped / By this mind and by that, / And every one is standard of his own chronology." For him it was intuitive.

"Your now is not my now," wrote Charles Lamb in England to his friend Barron Field in Australia, the far side of the earth, in 1817, "your then is not my then; but my now may be your then, and vice versa. Whose head is competent to these things?"

Nowadays we are all competent to these things. We have time zones. We can contemplate the International Date Line, where an imaginary boundary divides Tuesday from Wednesday.* Even when we suffer from jet lag—the quintessential disease of time travel—we are shrewd in our suffering and can nod wisely at William Gibson's account of "soul delay":

> Her mortal soul is leagues behind her, being reeled in on some ghostly umbilical down the vanished wake of the plane that brought her here, hundreds of thousands of feet above the Atlantic. Souls can't move that quickly, and are left behind, and must be awaited, upon arrival, like lost luggage.

We know that the light of the stars is ancient light, that distant galaxies reveal themselves to us only as they once were, not as they now are. As John Banville reminds us in his novel of that name, *ancient light* is all we have: "Even here, at this table, the light that is the image of my eyes takes time, a tiny time, infinitesimal, yet time, to reach your

* Time travel by circumnavigation? Poe seems to have been the first to make literary use of the possibility, in 1841 ("A Succession of Sundays," *Saturday Evening Post*), before Jules Verne made it the surprise ending of *Around the World in Eighty Days*.

eyes, and so it is that everywhere we look, everywhere, we are looking into the past."* (Can we peer into the future as well? That clever time traveler Joyce Carol Oates says via Twitter, "As minutes are required for the sun's light to reach us, we are living always in a sunlit past. Just the reverse, reading bound galleys.")

When everything reaching our senses comes from the past, when no observer lives in the now of any other observer, the distinction between past and future begins to decay. Events in our universe can be connected, such that one is the cause of the other, but, alternatively, they can be close enough in time and far enough apart that they cannot be connected and no one can even say which came first. (Outside the *light cone,* says the physicist.) We are more isolated, then, than we may have imagined, alone in our corners of spacetime. You know how fortune-tellers pretend to know the future? It turns out, said Richard Feynman, that no fortune-teller can even know the present.

Einstein's powerful ideas spread in the public press as rapidly as in the physics journals and disrupted the placid course of philosophy. The philosophers were surprised and outgunned. Bergson and Einstein clashed publicly in Paris and privately by post and seemed to be speaking different languages: one scientific, measured, practical; the

* The same thought came as a revelation to Israel Zangwill when he reviewed *The Time Machine* in 1895: "The star whose light reaches us to-night may have perished and become extinct a thousand years ago, the rays of light from it having so many millions of miles to travel that they have only just impinged upon our planet. Could we perceive clearly the incidents on its surface, we should be beholding the Past in the Present, and we could travel to any given year by travelling actually through space to the point at which the rays of that year would first strike upon our consciousness. In like manner the whole Past of the earth is still playing itself out—to an eye conceived as stationed to-day in space, and moving now forwards to catch the Middle Ages, now backwards to watch Nero fiddling over the burning of Rome."

other psychological, flowing, untrustworthy. "'The time of the universe' discovered by Einstein and 'the time of our lives' associated with Bergson spiraled down dangerously conflicting paths, splitting the century into two cultures," notes the science historian Jimena Canales. We are Einsteinian when we search for simplicity and truth, Bergsonian when we embrace uncertainty and flux. Bergson continued to place human consciousness at the center of time, while Einstein saw no place for spirit in a science that relied on clocks and light. "Time is for me that which is most real and necessary," wrote Bergson; "it is the necessary condition of action. What am I saying? It is action itself." Before an audience of intellectuals at the Société Française de Philosophie in April 1922, Einstein was adamant: "The time of the philosophers does not exist." Einstein, it seems, prevailed.

What does his framework mean for our understanding of the true nature of things? His biographer Jürgen Neffe sums up the situation judiciously. "Einstein provided no explanations for these phenomena," he says. "No one knows what light and time really are. We are not told *what* something is. The special theory of relativity merely provides a new rule for measuring the world—a perfectly logical construct that surmounts earlier contradictions."

HERMANN MINKOWSKI READ Einstein's 1905 paper on special relativity with special interest. He had been Einstein's mathematics teacher in Zurich. He was forty-four years old and Einstein was twenty-nine. Minkowski saw that Einstein had knocked the concept of time "from its high seat," had shown, indeed, that there is no *time,* but only *times.* But he thought that his former student had left the big job unfinished—had stopped short of stating the new truth about the nature of all reality. So Minkowski prepared a lecture. He delivered it at a scientific meeting in Cologne on September 21, 1908, and it is famous.

"Raum und Zeit" was his title, "Space and Time," and his mission was to declare both concepts null and void. "The views of space and time which I wish to lay before you have sprung from the soil of experimental physics, and therein lies their strength," he began grandly. "They are radical. *Henceforth space by itself, and time by itself, are doomed to fade away into mere shadows, and only a kind of union of the two will preserve an independent reality.*"

He reminded his listeners that space is denoted by three orthogonal coordinates, x, y, z, for length, breadth, and thickness. Let t denote time. With a piece of chalk, he said, he could draw four axes on the blackboard: "the somewhat greater abstraction associated with the number 4 does not hurt the mathematician." And so on. He was excited. This was "a new conception of space and time," he declared; "the first of all laws of nature." He called this conception the "principle of the absolute world."

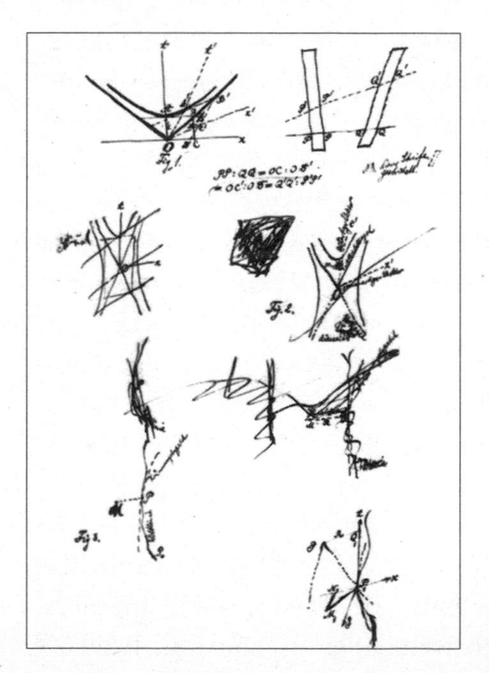

Four numbers, x, y, z, t, define a "world point." Together, all the world points that trace an object's existence from birth to death form a "world line." And what shall we call the whole shebang?

> The multiplicity of all thinkable *x, y, z, t* systems of values we will christen the *world.*

Die Welt! Good name. But we just call it spacetime now. (The continuum.) If we resist ("Because I know that time is always time/And place is always and only place," said T. S. Eliot), we do so in vain.

It was a bit of misdirection for Minkowski to begin by saying his lecture was grounded in experimental physics. His true subject was the power of abstract mathematics to reshape our understanding of the universe. He was a geometer above all. The physicist and historian Peter Galison puts it this way: "Where Einstein manipulated clocks, rods, light beams, and trains, Minkowski played with grids, surfaces, curves, and projections." He thought in terms of the most profound visual abstraction.

"Mere shadows," Minkowski said. That was not mere poetry. He meant it almost literally. Our perceived reality is a projection, like the shadows projected by the fire in Plato's cave. If the world—the absolute world—is a four-dimensional continuum, then all that we perceive at any instant is a slice of the whole. Our sense of time: an illusion. Nothing passes; nothing changes. The universe—the real universe, hidden from our blinkered sight—comprises the totality of these timeless, eternal world lines. "I would fain anticipate myself," said Minkowski in Cologne, "by saying that in my opinion physical laws might find their most perfect expression as reciprocal relations between these world lines." Three months later he was dead of a ruptured appendix.

Thus the idea of time as a fourth dimension crept forward. It did not happen all at once. In 1908 *Scientific American* "simply explained" the fourth dimension as a hypothetical space analogous to the first three: "For passing into the fourth dimension, we should pass out of our present world." The next year the magazine sponsored an essay contest on the topic "The Fourth Dimension," and not one of the winners or runners-up considered it to be time—notwithstanding the German physicists and the English writer of fantastic fiction. The space-time continuum was radical indeed. Max Wien, an experimental physicist, described his initial reaction as "a slight brain-shiver—now space and time appear conglomerated together in a gray, miserable chaos."* It offends common sense. "The texture of Space is not that of Time," cries Vladimir Nabokov, "and the piebald four-dimensional sport bred by relativists is a quadruped with one leg replaced by the ghost of a leg." If these critics sound Filbyish, even Einstein did not immediately embrace Minkowski's vision: *"überflüssige Gelehrsamkeit,"* he called it—superfluous learnedness. But Einstein came around. When his friend Besso died in 1955, Einstein consoled his family with words that have been quoted many times:

> Now he has departed from this strange world a little ahead of me. That means nothing. People like us, who believe in physics, know that the distinction between past, present, and future is only a stubbornly persistent illusion.

Einstein died three weeks later.

* Wien was the inventor of the *Löschfunkensender,* an early radio transmitter used, for example, on the *Titanic.*

FUNNY IRONY, though.

A century after Einstein discovered that perfect simultaneity is a chimera, the technology of our interconnected world relies on simultaneity as never before. When telephone-network switches get out of sync, they drop calls. While no physicist "believes in" absolute time, humanity has established a collective official timescale, preached by a choir of atomic clocks maintained at a temperature near absolute zero in vaults at the United States Naval Observatory in Washington, the Bureau International des Poids et Mesures near Paris, and elsewhere. They bounce their networked light-speed signals to one another, make the necessary relativistic corrections, and thus the world sets its myriad clocks. Confusion about past and future cannot be tolerated.

To Newton this would make perfect sense. International atomic time has the effect of codifying the absolute time that he created, and for the same reason: it lets the equations work out and the trains run on time. A century *before* Einstein, this technical achievement in simultaneity would have been almost impossible to conceive. The very notion of simultaneity scarcely existed. It was a rare philosopher who considered the question of what time it might be in a faraway place. One could hardly even hope to know, said the doctor and philosopher Thomas Browne in 1646,

> It being no ordinary or Almanack business, but a probleme Mathematical, to finde out the difference of hours in different places; nor do the wisest exactly satisfy themselves in all. For the hours of

several places anticipate each other, according to their Longitudes; which are not exactly discovered of every place.

All time was local. "Standard time" had no use before the railroad came and could not be established before the telegraph. England began *synchronizing its clocks* (new expression) to railway time in the mid-nineteenth century, when telegraph signals went out from the new electromagnetic clock at the Royal Observatory in Greenwich and the Electric Time Company in London. Also to the newly coordinated clock towers and electric street clocks of Bern.* These were technologies on which the ideas of Einstein depended, and also the ideas of H. G. Wells.

So now, on a hilltop near the Potomac River, the United States maintains a Directorate of Time, a subdepartment of the navy and by law the country's official timekeeper. Likewise in Paris is the BIPM, which also owns the international prototype of the kilogram. These are the keepers of *temps universel coordonné,* or coordinated universal time, or UTC—which I think we can admit is arrogantly named. Let's just call it Earth time.

All the chronometric paraphernalia of modernity: scientific, and yet arbitrary. Railroads made time zones inevitable, and in hindsight we can see that time zones already entailed a sense of time travel. They were not organized all at once, by fiat. They had many beginnings. For example, on November 18, 1883, a Sunday, known afterward as "the Day of Two Noons," James Hamblet, general superintendent

* Peter Galison, an authority on this matter, suggests that Einstein and Besso, conversing on that fateful day in May 1905, must have been standing on a hill in northeast Bern, where they could simultaneously see both Bern's old clock tower and another to the north, in the town of Muri.

of the Time Telegraph Company in New York City, reached out his hand and stopped the pendulum of the standard clock in the Western Union Telegraph Building. He waited for a signal and then restarted it. "His clock is adjusted to hundredth parts of a second," reported the *New York Times,* "a space of time so infinitesimal as to be almost beyond human perception." Around the city, tickers announced the new time and jewelers' shops adjusted their clocks. The newspaper explained the new setup in science-fictional terms:

> When the reader of The Times consults his paper at 8 o'clock this morning at his breakfast table it will be 9 o'clock in St. John, New-Brunswick, 7 o'clock in Chicago, or rather in St. Louis—for Chicago authorities have refused to adopt the standard time, perhaps because the Chicago meridian was not selected as the one on which all time must be based—6 o'clock in Denver, Col., and 5 o'clock in San Francisco. That is the whole story in a nut-shell.

Of course, that was nothing like the whole story. Arbitrary as they were, the railroads' time zones did not please everyone, and a new oddity followed: Daylight Saving Time, as it was known in North America, or, as Europeans called it, Summer Time. Even now, after a century of experience, some people find this twice-yearly time jump disturbing, and even physically uncomfortable. (And philosophically unsettling. Where does the hour go?) Germany was the first to impose *Sommerzeit,* during World War I, hoping to save coal. Soon after, the United States adopted it, then repealed it, then reimposed it. In England, seeking evening light for hunting, King Edward VII had the clocks on the royal estate set to "Sandringham Time," a half hour ahead of Greenwich. When the Nazis occupied France, they ordered all the clocks moved an hour forward, to Berlin time.

It was not just a matter of minutes and hours. The days and the years, too, confounded a world whose farthest parts were now in close communication. When, finally, would humanity agree on a uniform calendar? The new League of Nations took up the question after World War I. Its Committee on Intellectual Cooperation chose the philosopher Bergson as its president; another member, briefly, was Einstein. The League tried to impose the Gregorian calendar, itself the product of centuries of strife and revision, on nations less concerned with computing the proper dates of Easter feasts. The prospect of leaping forward or back in time created anxiety. Those nations did not fall in line. Bulgarians and Russians complained that their citizens could not be made suddenly to age by thirteen days, to surrender thirteen days of their lives in the name of globalization. Conversely, when France condescended to join Greenwich time, the Parisian astronomer Charles Nordmann said, "Some people may have consoled themselves with the reflection that to grow younger by 9 minutes and 21 seconds, on the authority of the law, was a pleasure worth having."

Had time become a thing over which dictators and kings could exercise power? "The Problem of Summer Time" is the English title of a new sort of time-travel story, published in 1943 by a darkly satirical Parisian writer, Marcel Aymé. In French it is *"Le décret"*—The Decree—issued after scientists and philosophers discover how easy it is to advance the time an hour forward each summer and back again each winter. "Little by little," says the narrator, "the realization spread that time was under man's control." Humans are time's dynamic masters: they may hurry up or slow down to suit themselves. Anyway "the old, stately pace was over."

> There was much talk of relative time, physiological time, subjective time and even compressible time. It became obvious that the notion of time, as our ancestors had transmitted it down the millennia, was in fact absurd claptrap.

With time now seemingly at their command, the authorities see a way to escape the nightmare of a war that seemed endless. They decide to advance the years by seventeen: 1942 leaps forward to 1959. (In the same spirit, moviemakers in Hollywood began to tear pages from calendars and spin the hands of clocks so as to move time along for their viewers.) The decree makes the world and all its people seventeen years older. The war has come to an end. Some have died, others have been born, and everyone has some catching up to do. It's all rather disorienting.

Aymé's narrator travels from Paris by railroad into the countryside. A surprise awaits him there. Apparently the decree has not spread everywhere. A storm, some wine, a troubled sleep, and in a distant village he encounters active German soldiers, and, sure enough, the mirror now shows him the thirty-nine-year-old he was, not the fifty-six-year-old he had become. On the other hand, he still has his newly acquired memories of those seventeen years. This is disturbing—indeed, impossible. "To be from an era, I thought, is to behold the world and oneself in a certain way that belongs to that era." Is he fated to relive the same life, burdened with memories of times to come?

He feels the existence of two parallel worlds, seventeen years apart yet existing simultaneously. Worse, after these "mysterious leaps and turns through time," why should there be only two?

> Now I accepted the nightmare of an infinity of universes, in which the official time represented only the relative displacement of my consciousness from one to another, and then on to another.

Now— and now— and then another now.

> Three o'clock—I am aware of the world in which I feature holding a pen. Three o'clock and one second—I am aware of the next universe in which I feature putting down my pen, etc.

It's too much for the human mind to comprehend; mercifully, his memories begin to fade, as all memories do. What he has written of the past—the future, and then the past—begins to seem like a dream. "Only once in a while, more and more infrequently, do I have the very ordinary sensation of déjà vu."

What is memory, for a time traveler? A conundrum. We say that memory "takes us back." Virginia Woolf called memory a seamstress "and a capricious one at that." ("Memory runs her needle in and out, up and down, hither and thither. We know not what comes next, or what follows after.")

"I can't remember things before they happen," says Alice, and the Queen retorts, "It's a poor sort of memory that only works backwards." Memory both is and is not our past. It is not recorded, as we sometimes imagine; it is made, and continually remade. If the time traveler meets herself, who remembers what, and when?

In the twenty-first century the paradoxes of memory grow more familiar. Steven Wright remarks: "Right now I'm having amnesia and déjà vu at the same time. I think I've forgotten this before."

FIVE

By Your Bootstraps

I don't want to talk about time travel shit, because we'll start talking about it and then we'll be here all day talking about it and making diagrams with straws.

—Rian Johnson (2012)

A MAN SITS in a locked room with his cigarettes, pots of coffee, and a typewriter. He knows all about time. He even knows about time travel. He is Bob Wilson, a Ph.D. candidate struggling to complete his thesis, "An Investigation into Certain Mathematical Aspects of a Rigor of Metaphysics." Case in point: "the concept 'Time Travel.'" He types, "Time travel may be imagined and its necessities may be formulated under any and all theories of time, formulae which resolve the paradoxes of each theory." More quasiphilosophical handwaving. "Duration is an attribute of consciousness and not of the plenum. It has no *Ding an Sich*."

Behind him he hears a voice. "Don't bother with it," the voice says. "It's a lot of utter hogwash anyway." Bob turns to see "a chap about the same size as himself and much the same age"—or maybe just a bit older, with a three-day beard and a black eye and a swollen upper lip. The chap has apparently emerged from a hole hanging in the air: "a great disk of nothing, of the color one sees when the eyes are shut tight." He opens a cupboard, finds the bottle, and helps himself to

Bob's gin. He looks vaguely familiar and he certainly knows his way around. "Just call me Joe," he says.

We see where this is going—we, people of the future, the time-savvy twenty-first century—but this story is taking place in 1941, and poor Bob is slow to catch on.

Bob's visitor explains that the hole in the air is a Time Gate. "Time flows along side by side on each side of the Gate. . . . You can walk into the future just by stepping through that circle." Joe wants Bob to walk through the Gate into the future. Bob doubts whether this is a good idea. As they discuss it, passing the gin bottle back and forth, a third man materializes. He bears a certain family resemblance to Bob and Joe. He does *not* want Bob to enter the gate. Now it's a committee. The phone rings: a fourth man, checking on everyone's progress.

Speculative philosophers and pulp readers had predicted this. In time travel, you can meet yourself. It's finally happening, and it's happening every which way. Before we are done we will have five protagonists, and they are all Bob. The author was Bob, too: Robert Anson Heinlein, writing under one of his several pen names, Anson MacDonald. His original title was "Bob's Busy Day"; the pulp magazine *Astounding Science Fiction* published it in October 1941 as "By His Bootstraps." It was the most intricate, complex, carefully plotted exercise in time travel to date.

No grandfathers die, no future mothers are impregnated, but wisecracks are exchanged and punches thrown. Scenes are narrated by one Bob and then reprised from the point of view of an older, more knowing Bob. You might expect "Joe" to remember his first encounter with Bob, but he finds the changed perspective confusing. Recognition only dawns slowly. The Bobs have to climb a ladder of growing self-awareness. To unravel the timeline we need a Minkowskiish diagram.

Heinlein drew one for himself while drafting the story.

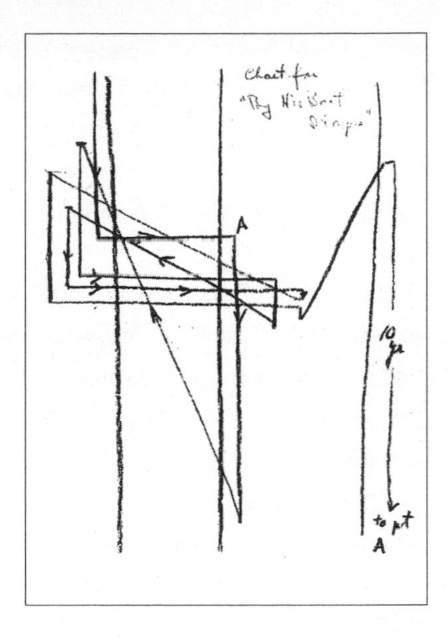

Really, of course, there are multiple timelines in play. Besides the Bobs', there is the reader's: the arc of the narrative. Our point of view is the one that matters. The author coaxes us gently along. He says of his poor hero, "He knew that he had about as much chance of understanding such problems as a collie has of understanding how dog food gets into cans."

Robert Heinlein came from Butler, Missouri, in the heart of the Bible Belt, and made his way to Southern California by way of the U.S. Navy, in which he served between the wars as a midshipman and sometime radio officer aboard the *Lexington,* one of the first aircraft carriers. He considered himself well skilled in ordnance and fire control, but after a collapse from pneumonia he was discharged as disabled. He wrote his first story in 1939 for a contest. *Astounding Science Fiction* paid him seventy dollars for it, and he began pounding the typewriter; he quickly became one of the pulps' most prolific and original writers. "By His Bootstraps" was one of more than twenty stories and short novels he published under various names in the next two years alone.

That first prizewinning story, "Life-Line," began in a familiar way: a mysterious man of science explains to a group of skeptical listeners that time is, still and always, the fourth dimension. "Maybe you believe it, perhaps not," he says. "It has been said so many times that it has ceased to have any meaning. It is simply a cliché that windbags use to impress fools." He asks them now to take it literally and to visualize the shape of a human being in four-dimensional spacetime. What is a human being? A spacetime entity measurable on four axes.

> In time, there stretches behind you more of this space-time event, reaching to, perhaps, 1905, of which we see a cross section here at right angles to the time axis, and as thick as the present. At the far end is a baby, smelling of sour milk and drooling its breakfast on its bib. At the other end lies, perhaps, an old man some place in the 1980s. Imagine this space-time event . . . as a long pink worm, continuous through the years.

A long pink worm. Slowly and gingerly, the culture was digesting the space-time continuum. The easy bits no longer needed quite so much explaining, so some nuances could be revealed.

The fun of "By His Bootstraps" lies in the comic encounters of the Bobs; it's a one-man farce times five, with misplaced hats, a confused and irate girlfriend (the word *two-timing* has never been so apt), and, with the Time Gate, the sci-fi equivalent of comically timed slamming doors. The hat is tossed and found and lost again until it seems to be multiplying like rabbits. Bob gets drunk with Bob. Bob recoils at the sight of drunken Bob, and Bob calls Bob some choice names. But Heinlein also takes some pains with the science. Or the philosophy. The eldest and wisest of the Bobs, living thirty thousand years in the future, tells one of his past selves, "Causation in a plenum need not be and is not limited by a man's perception of duration." Young Bob thinks about this and offers a comeback: "Just a second. How about entropy? You can't get around entropy." And so on. Examined closely, this gab is as hollow as the painted storefronts on the set of a Western.

Heinlein himself apparently didn't think much of the story at first and was surprised when the magazine's influential editor, John W. Campbell, assured him it was something special. In its way, it begins to grapple with two philosophical difficulties that arise when people start looping back through spacetime. One is the problem of who they are—the continuity of the self, let's call it. It's all well and good to talk about Bob Number One and Bob Number Two and so on, but the diligent narrator finds language ill equipped to keep everyone sorted: "His earlier self faced him, pointedly ignoring the presence of the third copy." Suddenly English doesn't have enough pronouns.

> His memory had not prepared him for who the third party would turn out to be.

> He opened his eyes to find that his other self, the drunk one, was addressing the latest edition.

Not only does Bob gaze upon himself—worse, he doesn't like the way he looks: "Wilson decided he did not like the chap's face." (But we don't need time travel to reproduce that experience. We have mirrors.)

What is the self? A question for the twentieth century to ponder, from Freud to Hofstadter and Dennett with detours through Lacan, and time travel provides some of the more profound variations on the theme. We have split personalities and alter egos galore. We have learned to doubt whether we *are* our younger selves, whether we will be *the same person* when we next look. The literature of time travel (though Bob Heinlein, in 1941, would not have dreamt of calling his work literature)* begins to offer a way into questions that might otherwise belong to philosophers. It looks at them viscerally and naïvely—as it were, nakedly.

If you're having a conversation with someone, can that person be you? When you reach out and touch someone, is it a different person, by definition? Can you have *memories* of a conversation while you're speaking the very words?

> Wilson's head started to ache again. "Don't do that," he pleaded. "Don't refer to that guy as if he were me. *This is me,* standing here."
>
> "Have it your own way. That is the man you *were.* You remember the things that are about to happen to him, don't you?"

He arrives at a conclusion: "The ego was himself. Self is self, an unproved and unprovable first statement, directly experienced." Henri Bergson would have appreciated this story.

> He thought of a way to state it: Ego is the point of consciousness, the latest term in a continuously expanding series along the line of memory duration. . . . He would have to try to formulate it mathematically before he could trust it. Verbal language had such queer booby traps in it.

He accepts the fact (because he remembers) that his earlier selves had also felt themselves to be the one and only integrated and continuous being, Bob Wilson. But that must be an illusion. In a four-

* When he writes of Bob Wilson, "His was a mixed nature, half hustler, half philosopher," Heinlein is proudly describing himself.

dimensional continuum each event is an absolute individual, with its own spacetime coordinates. "By sheer necessity he was forced to expand the principle of nonidentity—'Nothing is identical with anything else, not even with itself'—to include the ego. The Bob Wilson of now is *not* the Bob Wilson he had been ten minutes ago. Each was a discrete section of a four-dimensional process." All these Bobs—no more one and the same than the slices of bread in a loaf. And yet, they have continuity of memory, "a memory track that ran through all of them." He recalls something about Descartes. If we know anything about philosophy we know this: *cogito ergo sum.* We all feel that. It is the defining illusion of *Homo sapiens.*

As readers, how can we help but understand Bob as a unified self? We have lived with him through all the twists of his timeline. The self is the story he tells.

WE REACH (and it won't be the last time) the problem of free will. This was the second of the philosophical difficulties that Heinlein decided to explore as his narrative proceeded. Or perhaps I should say, found himself exploring, willy-nilly. He had no choice. When you send Bob back in time to meet his earlier self and relive an episode from his newer, wiser point of view, it is inevitable that Bob will ask: *Can't I do it differently this time?*

Then we loop again, and now Bob Three, older and wiser still, disagrees with Bob Two about what Bob One ought to do. He presumes that he, or they, have a choice. Will the earlier Bob defer to the superior wisdom of his later self? Hardly. He still needs to give one self a black eye and push the other self through the Time Gate.

The reader sees the whole picture—from above, so to speak—well before Bob does. Bob tries using the Time Gate as a window into spacetime, but the controls are hard to manage. Sometimes he sees, or senses, "flitting shadows which might be human beings." We know they are his own shadows, flickering on the cave wall. Bobs one and all are striving to fulfill their own destiny. The paradox, if it is a paradox, is that they have to work so hard, even as they gradually realize that their looping travails are foreordained. There is no escape from the track they are on. As Bob hears himself reciting words he has already spoken, he tries feebly to rewrite the script. "You're a free agent," he tells himself. "You want to recite a nursery rhyme—go ahead and do it . . . and break this vicious circle." Yet just at that moment he can't think of a nursery rhyme. His lines have been written for him. He can't get off the treadmill.

"But that's impossible!" he cries. "You're telling me that I did something because I was going to do something."

"Well, didn't you?" he calmly retorts. "You were there."

Young Bob still doesn't like it. "You would have me believe that causation can be completely circular." And Old Bob, despite all his hard-won knowledge, never stops working to fulfill his destiny. He does not wait for his earlier selves to play their roles; he manipulates them urgently. The narrator says: "Everyone makes plans to provide for their future. He was about to provide for his past." Taken all in all, this story is a snake pushing its own tail while musing about whether the effort is necessary.

The author, churning out stories on his manual typewriter to pay the bills in Southern California, trying to make his plots plausible and his characters convincing, has his own problem with free will. He makes his people into puppets, the strings flickering in and out of our sight. Their own view is foreshortened. Only the omniscient author, with his penciled diagrams, sees everything at once. We readers are caught up in the story, remembering the past, anticipating the future; we are mortals, for whom now means now.

It's not easy to get past that, in reading stories or in living our lives. As Heinlein puts it, we must make "a strong and subtle intellectual effort to think other than in durational terms, to take an eternal viewpoint." Free will cannot be easily dismissed, because we experience it directly. We make choices. No philosopher has yet sat down in a restaurant and told the waiter, "Just bring me whatever the universe has preordained." Then again, Einstein said that he could "will" himself to light his pipe without feeling particularly free. He liked to quote Schopenhauer: *Der Mensch kann wohl tun, was er will; aber er kann nicht wollen, was er will.* Man can do what he will, but he cannot will what he wills.

The free will problem was a sleeping giant and, without particularly meaning to, Einstein and Minkowski had prodded it awake. How literally were their followers to take the space-time continuum—the "block universe," fixed for eternity, with our blinkered three-dimensional consciousnesses moving through it? "Is the future all settled beforehand, and only waiting to be 'pushed through' into our three-dimensional ken?" asked Oliver Lodge, the British physicist and radio pioneer in 1920. "Is there no element of contingency? No free will?" He begged for a sort of modesty. "I am talking geometry, not theology, and it would be a stupid mistake to pretend to decide questions of high reality by aid of mere groping analogies and mathematical analysis. . . . The human race has not been in existence very long; it began its scientific studies very recently; it is still scraping on the surface of things, the three-dimensional surface of things." We may say the same, a century later.

PHILOSOPHERS DID NOT NEED the space-time continuum to tell them that there were problems about free will. As soon as the rules of logic were added to the human tool kit, the ancients found themselves capable of constructing the most amusing puzzles. Human language switches between past and future with a simple change of tense, and this can trap the unwary.

"For what is and what has come about, then, it is necessary that affirmation, or negation, should be true or false," Aristotle said. In other words, statements about the present and statements about the past are either true or false. Consider the proposition *There was a sea battle yesterday*. True or false. There is nothing in between. So it is natural to consider whether this applies to statements about the future. *There will be a sea battle tomorrow.* By Saturday this will be true or false, but must it be either true or false *now*? Put in terms of language and logic, these propositions look identical, so the same rules should apply. There will be a sea battle tomorrow. If it's not true or false, what else is there?

Aristotle remained unconvinced. He carved out an exception for propositions about the future. Where the future is concerned, he felt logic needed room for another state of things: call it indeterminate, contingent, unfixed, unknown, up for grabs . . . The modern philosopher finds this clumsy.

By the weekend, there *will have been* a sea battle. Not every language has a future perfect progressive tense built in; when your language does, it tends to feel natural. Either there will have been a sea battle or there won't. When the time comes, we'll know which. It will seem *to have been* inevitable. In this way, language and logic tend to suggest an eternalist view, the Universe Rigid, the view that gained solidity with the arrival of clockwork physical laws as revealed by Newton and Laplace. The block-universe package was wrapped and sealed, seemingly, in the four-dimensional space-time continuum. The new physics profoundly influenced philosophers, whether they acknowledged it or not. It freed them from the common intuitive sensation that past and future are quite different. It freed philosophers, that is, while imprisoning the rest of us. "Past and future must be acknowledged to be as real as the present," wrote Bertrand Russell in 1926, "and a certain emancipation from slavery to time is essential to philosophic thought."* A fatalist says: Everything that happens had to happen. Q.E.D.

Donald C. Williams, a realist from California, picked up that thread at midcentury with a paper on "The Sea Fight Tomorrow." His brand of realism was four-dimensional—fully modern, in other words. He asserted "the view of the world, or the manner of speaking about it" (a nice distinction, so easily forgotten),

> which treats the totality of being, of facts, or of events as spread out eternally in the dimension of time as well as the dimension of space. Future events and past events are by no means present events, but in a clear and important sense they do exist, now and forever, as rounded and definite articles of the world's furniture.

In the 1960s, the sea battle of tomorrow got a new life in the journals of philosophy. An argument raged over the logic of fatalism, and a

* "There is some sense, easier to feel than to state, in which time is an unimportant and superficial characteristic of reality."

milestone in the debate was the essay "Fatalism" by Richard Taylor, a metaphysician and beekeeper at Brown University. "A fatalist," he wrote, "thinks of the future in the manner in which we all think of the past." Fatalists take both past and future as given, and equally so. They may get this view from religion or, lately, from science:

> Without bringing God into the picture, one might suppose that everything happens in accordance with invariable laws, that whatever happens in the world at any future time is the only thing that can then happen, given that certain other things were happening just before, and that these, in turn, are the only things that can happen at that time, given the total state of the world just before then, and so on, so that again, there is nothing left for us to do about it.

Taylor proposed to prove fatalism entirely by philosophical reasoning, "without recourse to any theology or physics." He used symbolic logic, representing the various statements about the sea battle in terms of *P* and *P′* and *Q* and *Q′*. All he needed were "certain presuppositions made almost universally in contemporary philosophy." Something had to give: either fatalism or the rules of logic. A philosophy battle ensued. One of Taylor's presuppositions was not as evident to everyone else: "that time is not by itself 'efficacious'; that is, that the mere passage of time does not augment or diminish the capacities of anything." In other words, time itself is not an agent of change; more of an innocent bystander. Time doesn't do anything. ("What is a *mere* passage of time" retorted one of his critics. "Could time possibly pass without something, somewhere, changing—without the tick of a clock, the movement of a planet, the twitch of a muscle, or the sight of a flash?")

Two decades later, at Amherst College, an undergraduate philosophy student named David Foster Wallace, himself the son of a professional philosopher, grew obsessed with this nettlesome debate, "the famous and infamous Taylor argument." He wrote to a friend, "If you read the Taylor literature, it's really ulcer-city." He plunged in nonetheless. His obsession became his honors thesis, which might have taken its title from the imaginary Bob Wilson's "An Investigation into Certain Mathematical Aspects of a Rigor of Metaphysics." He drew diagrams to sort out "world-situations" and their possible "daughters" and "mothers." Yet as much as the formal, axiomatic side of philosophy appealed to Wallace—gave him continual pleasure and satisfaction—he never accepted it without reservation. The limits of logic and the limits of language remained live issues for him.

Words represent things but the words are not the things. We know that but we can forget. Fatalism is a philosophy built out of words, and ultimately its conclusions apply to *words*—not necessarily to reality. When Taylor leaves work, he summons the elevator just like the rest of us, by pressing the button. He does not think to himself, Don't worry, the elevator will follow its destiny. He may think, *When I press the elevator button, it is not a free choice—it was fated.* But he still goes to the trouble of doing it. He doesn't just stand there and wait.

Of course, Taylor himself knew this full well. He can't be refuted so easily.

> A fatalist—if there is any such—thinks he cannot do anything about the future. He thinks it is not up to him what is going to happen next year, tomorrow, or the very next moment. He thinks that even his own behavior is not in the least within his power, any more than the motions of the heavenly bodies, the events of remote history, or the political developments in China. It would, accordingly,

> be pointless for him to deliberate about what he is going to do, for a man deliberates only about such things as he believes are within his power to do.

He added, "And we are not, in fact, ever tempted to deliberate about what we have done and left undone."

I wonder whether Taylor had read much time-travel fiction or even, for that matter, whether he lived in the world I live in, where regret is not unknown and people do sometimes speculate about what might have been. Everywhere we look, people are pressing elevator buttons, turning doorknobs, hailing taxicabs, lifting sustenance to their lips, and begging their lovers' favor. We act as though the future is, if not in our control, not yet settled. Nonetheless, Taylor dismissed our "subjective feelings." We *would* suffer illusions of free will, because, by happenstance, we tend to know less about the future than about the past.

Many philosophers, in the years that followed, had tried to refute Taylor, but his logic proved amazingly robust. Wallace wanted to defend the common intuition "that persons as agents are capable of influencing the course of events in their world." He plunged into the depths of symbolic logic. "Since obviously under any analysis I have to do either O or O′ (since O′ is not-O), that is, since $(O \vee O')$; and since by (I-4) it is either not possible that I do O or not possible that I do O′, $(\sim\Diamond O \vee \sim\Diamond O')$, which is equivalent to $(\sim\Diamond\sim\sim O \vee \sim\Diamond\sim O)$, which is equivalent to $(\sim O \vee O)$, we are left with $(O \vee \sim O)$; so that it is necessary that whatever I do, O or O′, I do necessarily, and cannot do otherwise" is a sample sentence. (*"Obviously"*!) In the end he defeated Taylor's fatalism by stepping back and viewing not only the chains of symbols but also the levels of symbolic representation—viewing them, as it were, from above. Wallace distinguished between the realm of semantics and the realm of metaphysics. Considered strictly as words, he argued, Taylor's logic may be internally valid, but it's cheating to leap from semantic premises and arguments to a metaphysical conclusion.

"Taylor's claim was never really that fatalism was actually 'true,' only that it was forced upon us by proof from certain basic logical and semantic principles," he concluded. "If Taylor and the fatalists want to force upon us a metaphysical conclusion, they must do metaphysics, not semantics." In metaphysics we find the doctrine of determinism—we've seen this before, given its perfect expression by Laplace. Determinism is this (per Wallace):

> the idea that, given a precise and total state of affairs at one instant, and the physical laws that govern the causal relations between states of affairs, there is one and only one possible state of affairs that could obtain at the next instant.

Taylor takes this for granted. *If X, then Y* means one thing in logic. In the physical world, it means something trickier and always (we should know by now) subject to doubt. In logic, it is rigid. In physics, there is slippage. Chance has a part to play. Accidents can happen. Uncertainty is a principle. The world is more complex than any model.

Taylor was begging the question. To prove fatalism he was *assuming* determinism. Many physicists do that, too, even now. "Physicists like to think that all you have to do is say, 'These are the conditions, now what happens next,'" said Richard Feynman. Determinism is built into so many of their formalisms, just as it is for logicians. But formalisms are just that. The physical laws are a construct, a convenience. They are not coextensive with the universe.

Was that only possible which came to pass? Having spent years in these dark waters, Wallace had done enough philosophy for a while. He had an alternative future in mind, and he chose it. "I left there," he said later, "and I didn't go back."

SIX

Arrow of Time

The great thing about time is that it goes on. But this is an aspect of it which the physicist sometimes seems inclined to neglect.

—Arthur Eddington (1927)

WE ARE FREE to leap about in time—all this hard-won expertise must be good for something—but let's just set the clock to 1941 again. Two young Princeton physicists make an appointment to call at the white clapboard house at 112 Mercer Street, where they are led into Professor Einstein's study. The great man is wearing a sweater but no shirt, shoes but no socks. He listens politely as they describe a theory they are cooking up to describe particle interactions. Their theory is unconventional—full of paradoxes. It seems that particles must exert their influence on other particles not only forward in time but also backward.

John Archibald ("Johnny") Wheeler, thirty years old, had arrived at Princeton in 1938 after working with Niels Bohr in Copenhagen, at the citadel of the new quantum mechanics. Bohr had now sailed westward and Wheeler was working with him again, this time on the possibilities of nuclear fission in the uranium atom. Richard ("Dick") Feynman, age twenty-two, was Wheeler's favorite graduate student, a brash and whip-smart New Yorker. Johnny and Dick were nervous, and Einstein offered them sympathetic encouragement. He didn't mind the occasional paradox. He had considered something along these lines himself, back in 1909, as he recalled.

Physics is made of mathematics and words, always words and mathematics. Whether the words represent "real" entities is not always a productive question. In fact, physicists do well to ignore it. Are light waves "real"? Is the gravitational field? The space-time continuum? Leave it to theologians. One day the idea of *fields* is indispensable—you can practically feel them in your bones; anyway you can see the iron filings arranging themselves around the magnet—and the next day you wonder whether you can toss out fields and start over. That's what Wheeler and Feynman were doing. The magnetic field, also the electric field, but really just the electromagnetic field, was barely a century old, the invention (or discovery) of Faraday and Maxwell. Fields fill the universe: gravitational fields, boson fields, Yang-Mills fields. A field is a quantity that varies in space and time. It expresses variations in force. The earth feels the gravitational field of the sun, spreading outward through space. The apple dangling from the tree manifests the earth's gravitational field. Without fields, you have to believe in what looks like magic: action at a distance, through a vacuum, with no levers or strings.

Maxwell's equations for electromagnetic fields worked so beautifully, but by the 1930s and 1940s physicists were having problems in the quantum realm. They understood very well the equations connecting the energy of the electron with its radius. So they could compute the size of the electron quite precisely. Only, in quantum mechanics, it looks as though the electron has no radius at all: it is a point particle, zero-dimensional, taking up no space. Unfortunately for the mathematics, this picture led to infinities—the result of dividing by zero. To Feynman it seemed that many of these infinities came from a circular effect of the electron upon itself, its "self-energy." To eliminate these nasty infinities, he had the idea of simply not allowing electrons to act upon themselves. This meant eliminating the field. Particles would be allowed only to interact with other particles, directly. Not instantaneously: relativity had to be obeyed. The interactions occurred at the speed of light. That's what light *is:* interaction between electrons.

Feynman explained later, in Stockholm, upon receiving the Nobel Prize:

> It was just that when you shook one charge, another would shake later. There was a direct interaction between charges, albeit with a delay. The law of force connecting the motion of one charge with another would just involve a delay. Shake this one, that one shakes later. The sun atom shakes; my eye electron shakes eight minutes later, because of a direct interaction across.

The problem—if it was a problem—was that the rules for interaction worked backward in time as well as forward. They were symmetrical. This is the kind of thing that happens in Minkowski's world, where past and future are geometrically identical. Even before relativity, it was well known that Maxwell's equations for electromagnetism and, before that, Newton's for mechanics were symmetrical with respect to time. Wheeler had toyed with the idea that the positron—antiparticle of the electron—was an electron moving backward in time. So Johnny and Dick plunged ahead with a theory in which electrons appeared to be shining both forward into the future and back into the past. "I was enough of a physicist at that time," Feynman continued, "not to say, 'Oh, no, how could that be?' For today all physicists know from studying Einstein and Bohr that sometimes an idea which looks completely paradoxical at first, if analyzed to completion in all detail and in experimental situations, may, in fact, not be paradoxical."

In the end, the paradoxical ideas turned out not to be necessary for the theory of quantum electrodynamics. As Feynman well understood, such theories are models: never complete, never perfect, not to be confused with reality, which remains out of reach.

> It always seems odd to me that the fundamental laws of physics, when discovered, can appear in so many different forms that are not apparently identical at first, but, with a little mathematical fiddling you can show the relationship. . . . There is always another way to say the same thing that doesn't look at all like the way you said it before. . . .
>
> Many different physical ideas can describe the same physical reality.

On the side another issue loomed. Thermodynamics, the science of heat, offered a different version of time. Sure, the microscopic laws of physics say nothing about time having a favored direction. (Some would say "fundamental laws," rather than "microscopic laws," but that is not quite the same thing.) The laws of Newton, Maxwell, and Einstein are invariant with respect to past and future. Changing the direction of time is as easy as changing a sign from plus to minus. The microscopic laws are reversible. If you make a movie of a few colliding billiard balls or interacting particles, you can run the film through the projector backward and it will look fine. But make a movie of a cue ball breaking the rack—fifteen balls, at rest in a perfect triangle, sent flying to every corner of the table. If you play that one backward, it looks comically unreal: the balls careering about and then assembling themselves as if by magic into regimental order.

In the macroscopic world, the world we inhabit, time has a definite direction. When the technology of cinema was still new, filmmakers discovered they could create amusing effects by reversing their strips of celluloid. The Lumière brothers reversed their short *Charcuterie mécanique* to show a sausage unmade and a pig unbutchered. In a backward movie an omelet could organize into white and yolk and return to the egg, with shell fragments neatly reassembling themselves. A rock flies out of a turbulent pond, a reverse fountain of droplets closing in to seal the hole. Smoke pours down a fireplace into the flames as coals grow into logs. Not to mention life itself: the quintessential irreversible process. William Thomson, Lord Kelvin, saw the problem in 1874—and saw that consciousness and memory were part of the problem: "Living creatures would grow backward, with conscious knowledge of the future, but no memory of the past, and would become again unborn."

Every so often it is good to remind ourselves that most natural processes are *not* reversible. They work only one way, forward in time. For starters here is a little list from Lord Kelvin: "friction of solids; imperfect fluidity of fluids; imperfect elasticity of solids [all these *imperfects*]; inequalities of temperature, and consequent conduction of heat produced by stresses in solids and fluids; imperfect magnetic retentiveness; residual electric polarization of dielectrics; generation of heat by electric currents inducted by motion; diffusion of fluids, solutions of solids in fluids, and other chemical changes; and absorption of radiant heat and light." That last is where Johnny and Dick came in.

At some point we have to talk about entropy.

THERE'S A CATCHPHRASE, the *arrow of time,* familiarly used by scientists and philosophers in many languages (*la flèche du temps, Zeitpfeil, zamanın oku,* ось времени) as shorthand for a complex fact that everyone knows: time has a direction. The phrase spread widely in the 1940s and 1950s. It came from the pen of Arthur Eddington, the British astrophysicist who first championed Einstein. In a series of lectures at the University of Edinburgh in the winter of 1927 Eddington was attempting to comprehend the great changes under way in the nature of scientific thought. The next year he published his lectures as a popular book, *The Nature of the Physical World.*

It struck him that all previous physics was now seen to be *classical physics,* another new expression. "I am not sure that the phrase 'classical physics' has ever been closely defined," he told his listeners. No one called it classical until it broke down. (Now "classical physics" is a retronym, like acoustic guitar, dial telephone, and cloth diaper.)* Millennia had gone by without scientists needing special shorthand like "time's arrow" to state the obvious—*the great thing about time is that it goes on.* Now, however, it was no longer obvious. Physicists were writing laws of nature in a way that made time directionless, a mere change of sign separating $+t$ from $-t$. But one law of nature is different: the second law of thermodynamics. That's the one about entropy.

"Newton's equations go forwards and backwards, they do not care which way," explains Thomasina, the teenage prodigy invented by

* A retronym is a lexical time machine. It calls up entities past and present and juxtaposes them in the mind's eye.

Tom Stoppard in *Arcadia*. "But the heat equation cares very much, it goes only one way."

The universe tends inexorably toward disorder. Energy is indestructible, but it dissipates. This is not a microscopic law. Is it a "fundamental" law, like $F = ma$? Some argue that it is not. From one point of view, laws governing individual constituents of the world—single particles, or a very few—are primary, and laws about multitudes must be derived from them. But to Eddington this second law of thermodynamics was *the* fundamental law: the one that holds "the supreme position among the laws of Nature"; the one that gives us time.

In Minkowski's world past and future lie revealed before us like east and west. There are no one-way signs. So Eddington added one: "I shall use the phrase 'time's arrow' to express this one-way property of time which has no analogue in space." He noted three points of philosophical import:

1. It is vividly recognized by consciousness.
2. It is equally insisted upon by our reasoning faculty.
3. It makes no appearance in physical science except . . .

Except when we start to consider order and chaos, organization and randomness. The second law applies not to individual entities but to ensembles. The molecules in a box of gas comprise an ensemble. Entropy is a measure of their disorder. If you put a billion atoms of helium into one side of a box and a billion atoms of argon into the other side and allow them to bounce around for a while, they will not remain neatly separated but will eventually become a uniform—random—mixture. The probability that the next atom you find at a given place will be helium, rather than argon, will be 50 percent. The process of diffusion is not instantaneous and it runs in one direction. As you watch the distribution of the two elements, past and future are easily distinguishable. "A random element," said Eddington, "brings the irrevocable into the world." Without randomness, the clocks could run backward.

"The accidents of life" is the way Feynman liked to put it: "Well, you see that all there is to it is that the irreversibility is caused by the general accidents of life." If you throw a cup of water into the sea, let time pass, and dip your cup back in, can you get the same water back? Well, you could—the probability is not zero. It's just awfully small. Fifteen billiard balls could smash around a table and finally come to a stop in a perfect triangle—but when you see that happen, you know that the film has been reversed. The second law is a probabilistic law.

Mixing is one of those processes that follow the arrow of time. Unmixing takes work. "You cannot stir things apart," says Stoppard's Thomasina—entropy explained in five words. (Her tutor, Septimus, replies, "No more you can, time must needs run backward, and since it will not, we must stir our way onward mixing as we go, disorder out of disorder into disorder until pink is complete, unchanging and unchangeable, and we are done with it for ever.") Maxwell himself wrote:

> *Moral.* The 2nd law of Thermodynamics has the same degree of truth as the statement that if you throw a tumblerful of water into the sea, you cannot get the same tumblerful of water out again.

But Maxwell predated Einstein. For him, time required no particular justification. He already "knew" that the past is past and the future still to come. Now matters are not so simple. In 1949, in a essay titled "Life, Thermodynamics, and Cybernetics," Léon Brillouin said:

> Time flows on, never comes back. When the physicist is confronted with this fact he is greatly disturbed.

To the physicist, it feels that a troublesome gap lies between the microscopic laws, where time has no preferred direction, because the laws are reversible, and the macroscopic world, where the arrow of time points from past to future. Some are content to say that fundamental processes are reversible and macro-scale processes are mere statistics. This gap is a disconnect—a lapse in explanation. How do you get from one place to the other? The gap even has a name: the arrow of time dilemma, or Loschmidt's paradox.

Einstein admitted that the problem disturbed him at his moment of greatest understanding, in the creation of the general theory of relativity—"without my having succeeded in clarifying it." In a diagram of the four-dimensional space-time continuum, let's say that *P* is a "world-point" lying between two other world-points, *A* and *B*. "We draw a 'time-like' world-line through *P*," suggested Einstein; "does it make any sense to provide the world-line with an arrow, and to assert that *B* is before *P*, *A* after *P*?" Only when thermodynamics enters the picture, he concluded—but he also said that *any transfer of information* involves thermodynamics. Communication and memory are entropic processes. "If it is possible to send (to telegraph) a signal from *B* to *A*, but not from *A* to *B*, then the one-sided (asymmetrical) character of time is secured, i.e. there exists no free choice for the direction of the arrow. What is essential in this is the fact that the sending of a signal is, in the sense of thermodynamics, an irreversible process, a process which is connected with the growth of entropy."

In the beginning, therefore, the universe must have had low entropy. *Very* low entropy. It must have been in a highly ordered state, which is also an extremely improbable state. This is a cosmic mystery. Ever since, entropy has grown. "That is the way toward the future," said Feynman, years later, when he was a famous man assembling his knowledge of physics into textbook form.

> That is the origin of all irreversibility, that is what makes the processes of growth and decay, that makes us remember the past and not the future, remember the things which are closer to that moment in history of the universe when the order was higher than now, and why we are not able to remember things where the disorder is higher than now, which we call the future.

And in the end?

THE UNIVERSE TENDS toward maximum entropy, the condition of ultimate disorder from which there is no return. The eggs will all have scrambled, the sand castles blown down, the sun and stars faded to uniformity. H. G. Wells already knew about entropy and heat death. This is the destiny the Time Traveller nears, when he abandons Weena, departs the year 802,701, leaves behind the troglodytic Eloi and bovine Morlocks, the ruined Palace of Green Porcelain, its Gallery of Palaeontology long deserted, its library a wilderness of rotting paper, and drives his machine onward, swaying and vibrating through millions of years of grayness into a final twilight brooding over the earth. If you read *The Time Machine* when you are young, I think this is what lodges in your memory or your dreams, this final tableau where nothing happens. In one draft Wells called it "The Further Vision." If Eden is alpha, here is omega. Eschatology for the enlightened. No hell, no apocalypse. Not with a bang but a whimper.

This twilight beach recurs again and again in science fiction. We come to land's end—J. G. Ballard's "derelict landscape," the terminal beach, where the last man says farewell: "Such a leave-taking required him to fix his signature on every one of the particles in the universe." In Wells's unforgettable final pages, the Time Traveller sits shivering in his saddle and watching "the life of the old earth ebb away." Nothing stirs. All he sees is stained red, pinkish, bloody, in the dim light of the dying sun. He imagines some black thing flopping about, but it is only a rock.

> I stared aghast at the blackness that was creeping over the day. . . . A cold wind began to blow. . . . Silent? It would be hard to convey the stillness of it. . . . The darkness thickened. . . . All else was rayless obscurity. . . . A horror of this great darkness came on me. The cold, that smote my marrow.

This is the way the world ends.

SEVEN

A River, a Path, a Maze

Time is a river that sweeps me along, but I am the river; it is a tiger that mangles me, but I am the tiger; it is a fire that consumes me, but I am the fire.

—Jorge Luis Borges (1946)

TIME IS a river. Does the truism require elaboration?

It did in 1850. Case in point: an American novel titled *The Mistake of a Life-Time; or, The Robber of the Rhine Valley. A Story of the Mysteries of the Shore, and the Vicissitudes of the Sea.* The author, Waldo Howard, promises "a truthful panorama of the events of a stirring and romantic period." Let us jump to chapter 13, "Lady Gustine and the Jew."

Lady Gustine is a dignified and high-toned beauty of eighteen years ("summers"), while her companion for the evening (not the Jew, obviously) is an equally dignified and beautiful twenty-year-old. They have been dancing. She is fatigued. "I fear you are fatigued," says the gentleman. " 'Oh, no,' said the lady, panting to regain the breath she had expended in the waltz."

Conveniently, their balcony overlooks a river. They gaze upon it awhile. Presently dialogue occurs:

"Are you dreaming?"

"O, no, lady. I—I was thinking how truly the passage of yonder tiny craft resembles that of our own life bark on the tide of time."

"And how?"

"See you not how quietly its hull is borne along with the current? . . . [etc., etc.]

"Well." [He's boring her.]

"Thus we are moving now, lady, rapidly, with silent, but steady, and never ceasing motion, down the swift river of time, that sets through the valley of life; all unconsciously we glide on, nodding like this same helmsman, indifferently, as we hold the rudder that guides our own fate—while we swiftly approach the ocean of eternity."

And more like that. Pretty soon he "dwells upon the beauties of her native valley" but we needn't follow him there. The first metaphor is bad enough.

Time = river. Self = boat. Eternity = ocean.

When time is a river, then time travel becomes plausible. You might get out and run up or down the banks.

People have been comparing time to a river at least since Plato began a long tradition of misquoting Heraclitus: "You can't step into the same river twice." Or "We step and do not step into the same rivers." Or "We both step and do not step, are and are not in the same rivers."* No one knows exactly what Heraclitus said, because he lived in a time and place that lacked writing (his work is published under the title *The Complete Fragments,* no irony intended), but according to Plato:

* To the extent that Heraclitus's actual words can be reconstructed and translated into English, another version is this: *On those stepping into rivers staying the same other and other waters flow.*

> Heraclitus, I believe, says that all things pass and nothing stays, and comparing existing things to the flow of a river, he says you could not step twice into the same river.

Heraclitus was saying something important: namely, that things change. The world is in flux. That may seem self-evident, but his approximate contemporary Parmenides took a different view: change is an illusion of our senses; beneath the transitory world of appearances lies the true reality—stable, timeless, eternal. This was the view that appealed to Plato.

Notice that no one so far is saying that time is like a river. The universe is like a river. It flows. (Or it doesn't, if you're Plato.)

Alfred Jarry, constructing his time machine in 1899, said it had already become "a banal poetic figure to compare Time to a flowing stream."* Banality didn't stop anyone. "Time, that impalpable and fatal river," the Parisian astronomer Charles Nordmann said in 1924, "strewn with dead leaves, our wistful hours carried down stream." Where are we in this picture—we, the conscious observer? We are merely a bump in the viscosity, said the absurdist Jarry. The Christian hymn says, "Time, like an ever rolling stream / Bears all its sons away." The river carries us toward eternity, which is to say past death. Miguel de Unamuno wrote, *"Nocturno el río de las horas fluye . . ."*—though he imagined it flowing *from* the future, *"el mañana eterno."* Marcus

* Nabokov took the same jaundiced view a century later: "We regard Time as a kind of stream, having little to do with an actual mountain torrent showing white against a black cliff or a dull-colored great river in a windy valley, but running invariably through our chronographical landscapes. We are so used to that mythical spectacle, so keen upon liquefying every lap of life, that we end up by being unable to speak of Time without speaking of physical motion."

Aurelius, Stoic philosopher and emperor, said that time is a river because everything rushes by, while we watch. "No sooner has anything appeared than it is past, and now another thing is passing, and that yonder will presently be here."

If time is a river, can we ask how fast it flows? That seems a natural question to ask about a river, but it's not a good question to ask about time itself. How fast does time flow? Measured how? We have plunged into a tautology. It's no better to ask, How fast are we advancing through time?

Riverine flow can be complicated. Can temporal flow? "There is a theory," explains Spock in a classic episode of *Star Trek*. "There could be some logic to the belief that time is fluid, like a river, with currents, eddies, backwash."

If time is a river, does it have tributaries? Whence does it spring? From the big bang, or are we now mixing metaphors? If time is a river, where are the banks that contain it? W. G. Sebald asked that question in his last novel, *Austerlitz:*

> Where, seen in those terms, where are the banks of time? What would be this river's qualities, qualities perhaps corresponding to those of water, which is fluid, rather heavy, and translucent?

Sebald also asked, "In what way do objects immersed in time differ from those left untouched by it?" This was a nice conceit: that some parts of our world, like dusty, shuttered rooms, may stand outside of time, may be cut off from time, immune to the flow.

IN POINT OF FACT, time is not a river. We possess a great metaphorical tool kit with utensils for every occasion. We say that time *passes,* time *goes by,* and time *flows,* and all those are metaphors. "Time is a fluid medium for the culture of metaphors," writes Nabokov metaphorically. We also think of time as a medium in which we exist. And as a quantity that we can *possess, waste,* or *save.* Time is like money, it is like a road, a path, a maze (Borges again, of course), a thread, a tide, a ladder, and an arrow. All at once.

"The idea that Time 'flows' as naturally as an apple thuds down on a garden table implies that it flows in and through something else," says Nabokov, "and if we take that 'something' to be Space then we have only a metaphor flowing along a yardstick."

Is it even possible to talk about time without using metaphors? Perhaps:

> Time present and time past
> Are both perhaps present in time future,
> And time future contained in time past.

Although, if that isn't metaphor, what trope is it? Pregnant words: "present in . . ."; "contained in . . ." In the same poem T. S. Eliot also had some words about words.

> Words strain,
> Crack and sometimes break, under the burden,
> Under the tension, slip, slide, perish,
> Decay with imprecision, will not stay in place,
> Will not stay still.

Everything was so unsettled about time. Philosophers, physicists, poets, and pulp writers all struggled. They were using the same word bag. They drew their tiles and moved them around the playing board. (Slip, slide, perish, decay with imprecision.) Philosophers' words alluded to the philosophers' words that came before. Physicists' words were special, more precisely defined, and anyway they were mostly numbers. Physicists don't generally call time a river. They don't generally depend on metaphor; at least, they don't like to admit it. Even "arrow of time" is not so much a metaphor as a catchphrase.

In the twentieth century the physicists took the moral lead—they had the power—and the philosophers mainly reacted or resisted. After Einstein's message sank in, metaphysicians began to say without blushing that time and space have the same "ontological standing," that they exist "in the same way." As for poets, they lived in the same world, pulled the same tiles from the bag, and knew better than to trust all the words. Proust searching for lost time. Woolf stretching and warping it. Joyce assimilating the news about time as it came from the frontier of science. "Temporal or spatial," says Stephen in *A Portrait of the Artist as a Young Man,* "the esthetic image is first luminously apprehended as selfbounded and selfcontained upon the immeasurable background of space or time which is not it." No, it is not. Later came *Ulysses,* the book of a single day, exodus and return. "An unsatisfactory equation between an exodus and return in time through reversible space and an exodus and return in space through irreversible time." Leopold Bloom worries about magnetism and time, the sun and the stars, pulling and being pulled: "Very strange about my watch. Wristwatches are always going wrong." Oh, there was unease.

Not everyone liked T. S. Eliot's last long poem, *Four Quartets,* published from 1936 to 1942. Some accused it of self-parodic inscrutability. Not everyone thought it was a poem about time, but it is. *Here the impossible union / Of spheres of existence is actual, / Here the past and future / Are conquered, and reconciled.* Does all time exist together? Is the future already contained in the past? Didn't Einstein say so?

Along with quite a few of his contemporaries, Eliot was influenced by a slightly crackpot book, *An Experiment with Time,* written by an Irish aeronautical pioneer named John William Dunne. Dunne was an acquaintance of Wells who at the turn of the century began building aircraft models, then gliders, then powered biplanes, all tailless (a design with stability problems). In the twenties, having left aeronautics behind, he noticed that his dreams sometimes predicted future events. They were "precognitive dreams," he decided. Reverse memory. He had dreamed of a volcano killing four thousand on a French island and then, later (or so he recalled), read in the newspaper of the Pelée eruption on Martinique, killing forty thousand. He began keeping a notebook and pencil under his pillow; he interviewed his friends about their dreams; and he put two and two together. By 1927 he had a theory and a book.

Dunne proposed to replace the foundations of epistemology with his new system. "If prevision be a fact, it is a fact which destroys the entire basis of all our past opinions of the universe." The past and the future coexist, in "the time dimension." Incidentally, he wrote, he had stumbled upon "the first scientific argument for human immortality." He put forward not a four-dimensional but a five-dimensional view of space and time. In explaining this, he adverted to Einstein and Minkowski and, as another authority, to Mr. H. G. Wells, who "through the mouth of one of his fictional characters, stated his case with a clearness and conciseness which has rarely, if ever, been surpassed."

Wells himself did not approve. He assured Dunne that "prevision" was claptrap and that time traveling was make-believe—"that I [Dunne] have taken something which he never intended to be treated seriously . . . and have brooded too much upon it." But Eliot and other literary searchers absorbed Dunne's provocative ideas and imagery, including the prospect of a kind of immortality. *The future is a faded song,* Eliot writes. *The way up is the way down* (another fragment from Heraclitus), *and the way forward is the way back.* He has sensed that all time is eternally present, but he is not sure.* *If all time is eternally present / All time is unredeemable.*

The Universe Rigid? Eliot in *Four Quartets* is not trying to persuade us of a system of the world. He suffers paradox and self-doubt. "I can only say, *there* we have been: but I cannot say where. / And I cannot say, how long, for that is to place it in time." He speaks through masks. Not only are words slippery; the problem with using words to describe time is that words themselves are *in* time. A string of words has a beginning, a middle, and an end. "Words move, music moves / Only in time." Is eternity a place of motion or of stillness? Movement or pattern? Can these coexist? *At the still point of the turning world?* When he says a Chinese jar moves perpetually in its stillness, you know that's a metonym. What moves perpetually in its stillness is a poem.†

* Seeing a photograph album in 1917, he wrote to his mother, "It gives one the feeling that Time is not before and after, but all at once, present and future and all the periods of the past, an album like this."

† Only by the form, the pattern,
Can words or music reach
The stillness, as a Chinese jar still
Moves perpetually in its stillness.

You shall not think "the past is finished" or "the future is before us." Time does not belong to us; we cannot grasp it or define it. We can barely count it. *The tolling bell,* Eliot tells us,

> Measures time not our time, run by the unhurried
> Ground swell, a time
> Older than the time of chronometers, older
> Than time counted by anxious worried women
> Lying awake, calculating the future,
> Trying to unweave, unwind, unravel
> And piece together the past and the future,
> Between midnight and dawn, when the past is all deception,
> The future futureless.

WHEN BORGES, the philosopher poet, wrote that time is a river, he meant approximately the opposite. Time is not a river, nor is it a tiger, nor a fire. Borges, the critic, used a bit less paradox, a bit less misdirection. His language regarding time is apparently plain. In 1940 he, too, wrote about Dunne and his *Experiment with Time,* declaring it absurd, in a mild way. Part of Dunne's argument was a reflection on consciousness—how it cannot be contemplated without falling into recursive loops ("a conscious subject is conscious not only of what

---

Not the stillness of the violin, while the note lasts,
Not that only, but the co-existence,
Or say that the end precedes the beginning,
And the end and the beginning were always there
Before the beginning and after the end.
And all is always now.

it observes, but of a subject A that also observes and therefore, of another subject B that is conscious of A, and . . ." on and on). He was onto something important, recursion as an essential feature of consciousness, but then he concluded that "these innumerable intimate observers do not fit into the three dimensions of space, but they do in the no less numerous dimensions of time." Borges knew this was nonsense, and it was his kind of nonsense. He saw something in it, a way to think about how the perception of time must be built on memory: "successive (or imaginary) states of the initial subject." He recalled an observation made by Gottfried Wilhelm Leibniz: "If the spirit had to reflect on each thought, the mere perception of a sensation would cause it to think of the sensation and then to think of the thought and then of the thought of the thought, and so to infinity." We create memories or our memories create themselves. Consulting a memory converts it into a memory of a memory. The memories of memories, the thoughts of thoughts, blend into one another until we cannot tease them apart. Memory is recursive and self-referential. Mirrors. Mazes.*

Dunne's precognitive dreams and involuted logic led him to a belief in a preexisting future, an eternity within human reach. Borges said Dunne was making the mistake "those absentminded poets" make when they start to believe their own metaphors. By absentminded poets he seemed to mean physicists. By 1940 the new physics took the fourth dimension and the space-time continuum as real, but Borges emphatically did not:

* And corridors. "When we remember our former selves, there is always that little figure with its long shadow stopping like an uncertain belated visitor on a lighted threshold at the far end of an impeccably narrowing corridor." —Vladimir Nabokov, *Ada, or Ardor.*

> Dunne is an illustrious victim of that bad intellectual habit—denounced by Bergson—of conceiving time as a fourth dimension of space. He postulates that the future toward which we must move already exists (also conceived in spatial form, in the form of a line or a river).*

Borges had more to say than most about the problem of time in the twentieth century. For him paradox was not a problem but a strategy. He *believed* in time—its reality, its centrality—yet he titled his crucial essay "A New Refutation of Time." Of eternity he was not so fond. In another essay, "A History of Eternity," he declared: "For us, time is a jarring, urgent problem, perhaps the most vital problem of metaphysics, while eternity is a game or a spent hope." Everyone "knows" (said Borges) that eternity is the archetype and our time merely its fleeting image. He proposed the opposite: Time comes first; eternity is created in our minds. Time is the substance, eternity the effigy. Contrary to Plato—contrary to the Church—eternity is "more impoverished than the world." If you are a scientist, you may substitute infinity. That is your creation, after all.

As for his new refutation of time, its essence is an argument he has "glimpsed" or "foreseen" and in which he himself does not believe. Or does he? It comes to him in the night. In the Proustian hours. What is time when you awaken, between dreams, register the rustling sounds, the shadowy walls—or, let's say you're Huckleberry Finn, rafting down the *river* . . .

* Nor, by the way, did Borges express great love for Eliot. "You always think—at least I always feel—that he's agreeing with some professor or slightly disagreeing with another." He accused him of a rather subtle form of humbug: "the deliberate manipulation of anachronisms to produce an appearance of eternity."

> Negligently he opens his eye: he sees an indefinite number of stars, a nebulous line of trees. Then he sinks into a sleep without memories, as into dark waters.

Borges notes that this is "a literary, not a historical" case. The doubting reader is invited to substitute a personal memory. Think of an incident in your past. *When* is that memory? Not at any time—not at any *precise* time. It is an instant on its own, suspended, apart from any supposed space-time continuum. Spacetime? "I tend to be always thinking of time, not of space," Borges writes. "When I hear the words 'time' and 'space' used together, I feel as Nietzsche felt when he heard people talking about Goethe and Schiller—a kind of blasphemy."

He denies simultaneity, just as Einstein did, only Borges does not care about the signal velocity (light speed) because our natural state is alone and autonomous, our signals are fewer and less reliable than the physicist's.

> The lover who thinks, "While I was so happy, thinking about the faithfulness of my beloved, she was busy deceiving me," is deceiving himself. If every state in which we live is absolute, that happiness was not concurrent with that betrayal.

The lover's knowledge cannot modify the past, though it can modify the recollection. Having dispensed with simultaneity, Borges also denies succession. The continuity of time—the whole of time—another illusion. Furthermore, this illusion, or this problem, the never-ending effort to assemble a whole from a succession of instants, is also the problem of identity. Are you the same person you used to be? How would you know? Events stand alone; the totality of all events is an idealization as false as the sum of all the horses: "The universe, the sum total of all events, is no less ideal than the sum of all the horses—one, many, none?—Shakespeare dreamed between 1592 and 1594." Oh, Marquis de Laplace.

We have a tendency to take our words too seriously, which happens (paradoxically) when we are unconscious of them. Language offers a woefully meager set of choices for expressing what we need to express. Consider this sentence: "I haven't seen you for a [?] time." Must the missing word be *long*?* Then time is like a line or a distance—a measurable space. The language forces this upon us. Who was the first person to say that time "passes" or time "flows"? We are seldom conscious of the effect of language on our choice of metaphors, the effect of our metaphors on our sense of reality. Usually we give the words no thought at all. When we do, we may well wonder what we're really saying. "I'm terrified of the thought of time passing (or whatever is meant by that phrase) whether I 'do' anything or not," Philip Larkin wrote to his lover Monica Jones. The words lead us in a certain direction.

In English and most Western languages, the future lies ahead. In front of us. Forward. The past is behind us, and when we are running late we say we have fallen behind. Yet this forward-backward orientation is neither obvious nor universal. Even in English, it seems we can't agree on what it means to move a meeting *back* one day. Some people are certain that back means earlier. Others are equally certain that it means later. On Tuesday, Wednesday lies before us, though Tuesday is before Wednesday. Other cultures have different geometries. Aymara speakers, in the Andes, point forward (where they can

* In English "long" is almost forced; in other languages, that would sound bizarre. They might say "big."

see) when talking about the past and gesture behind their backs when talking about the future. In other languages, too, yesterday is the day ahead and tomorrow is the day behind. The cognitive scientist Lera Boroditsky, a student of spatiotemporal metaphors and conceptual schemas, notes that some Australian aboriginal communities orient themselves by cardinal direction (north, south, east, west) rather than relative direction (left, right) and think of time as running east to west. (They have a strongly developed sense of direction, compared to more urban and indoor cultures.) Mandarin speakers often use vertical metaphors for time: 上 (*shàng*) means both above and earlier; 下 (*xià*) means below and next. The up month is the one that just ended. The down month is on its way.

Or are *we* on *our* way? Boroditsky and others speak in terms of "ego-moving" versus "time-moving" metaphors. One person may feel the deadline approaching. Another may feel herself approaching the deadline. These may be the same person. You may swim onward, or the river may bear you.

If time is a river, are we standing on the bank or bobbing along? "To say time passes more quickly, or that time flows, is to imagine *something* flowing," wrote Wittgenstein.

> We then extend the simile and talk about the direction of time. When people talk of the direction of time, precisely the analogy of a river is before them. Of course a river can change its direction of flow, but one has a feeling of giddiness when one talks of time being reversed.

That is the giddiness of the time traveler—like looking at an Escher staircase. Time *passes.* "The hours pass slowly." "The hours pass quickly." And without contradicting ourselves, we pass the time. We say these words, and we understand them perfectly.

Time is not a river. Where does that leave time travel?

A MAN LIES supine on an iron cot in a locked room, pondering his own imminent death. Through the window he can see roofs and the sun, shaded by clouds. He is aware of the time: it is a "six o'clock sun." His name may or may not be Yu Tsun. We gather that he is a German spy. He is in possession of the Secret. The Secret is a single word, a name, "the exact location of the British artillery park on the River Ancre." But he has been discovered and marked for assassination. He turns out to be something of a philosopher.

> It seemed incredible to me that that day without premonitions or symbols should be the one of my inexorable death. . . . Then I reflected that everything happens to a man precisely, precisely *now*. Centuries of centuries and only in the present do things happen; countless men in the air, on the face of the earth and the sea, and all that really is happening is happening to me.

This is a fiction by Borges, "*El jardín de senderos que se bifurcan,*" the title story of his first collection—eight stories, sixty pages—published in 1941 by the modernist journal *Sur* in Buenos Aires. Borges, who read *The Time Machine* with excitement when he was young, had published some poetry and some criticism. He was a prolific translator from English, French, and German, including Poe, Kafka, Whitman, and Woolf. To support himself he worked as an assistant at a small, down-at-the-heels branch library, cataloguing and cleaning the books.

Seven years later, "The Garden of Forking Paths" became Borges's first story to appear in English translation. His American publisher was not a literary establishment or journal but *Ellery Queen's Mystery Magazine,* August 1948. He did appreciate mystery. Now his reputation is large, but he did not gain much fame in English-speaking countries until the sixties, when he shared the first Prix International with Samuel Beckett. By then he was an old man, and blind.

Ellery Queen (joint pseudonym for two cousins from Brooklyn) was happy to publish what could barely be called a detective story. It has no detective, but it does have a struggle among spies, a pursuit, a revolver chambering a single bullet, a confrontation, and a murder. There is not just a mystery but a philosophical mystery—so we are told. Yu Tsun is informed, "Philosophic controversy usurps a good part of the novel." To what does the controversy pertain?

> I know that of all problems, none disturbed him so much as the abysmal problem of time. Now then, the latter is the only problem that does not figure in the pages of the *Garden* . . .
>
> *The Garden of Forking Paths* is an enormous riddle, or parable, whose theme is time; this recondite cause prohibits its mention.

The story folds in upon itself: *The Garden of Forking Paths* is a book inside a book. (And now inside a pulp magazine.) The *Garden* is a meandering novel by "the oblique Ts'ui Pên." It is a book that is also a maze. It is a set of chaotic manuscripts, "an indeterminate heap of contradictory drafts." It is a labyrinth of symbols. It is a labyrinth of time. It is infinite—but how can a book, or a maze, be infinite? The book says, "I leave to the various futures (not to all) my garden of forking paths."

The paths fork in time, not in space.

> *The Garden of Forking Paths* is an incomplete, but not false, image of the universe as Ts'ui Pên conceived it. In contrast to Newton and Schopenhauer, your ancestor did not believe in a uniform, absolute time. He believed in an infinite series of times, in a growing, dizzying net of divergent, convergent and parallel times. This network of times which approached one another, forked, broke off, or were unaware of one another for centuries, embraces *all* possibilities of time.

In this, as in so many things, Borges seemed to be peering over the horizon.* Later the literature of time travel expanded to encompass alternative histories, parallel universes, and branching time lines. A parallel adventure was under way in physics. Having drilled far down inside the atom, to a place where particles are inconceivably small and behave sometimes like particles and sometimes like waves, physicists encountered what appears to be an inescapable randomness at the heart of things. They were continuing the project of computing future states from specified initial conditions at time $t = 0$. Only now they were using wave functions. They were solving the Schrödinger equation. Calculations of wave functions via the Schrödinger equation produce not specific results but probability distributions. You may remember Schrödinger's cat: either alive or dead, or neither alive nor

* Even before Borges, a twenty-year-old in Colorado named David Daniels wrote a story for *Wonder Stories* in 1935 called "The Branches of Time": a man with a time machine discovers that when he returns to the past, the universe splits into parallel world lines, each with its own history. The next year, Daniels killed himself with a gun.

dead, or, if one prefers (it's something of a matter of taste), simultaneously alive and dead. Its fate is a probability distribution.

When Borges was forty years old and writing "The Garden of Forking Paths," a boy named Hugh Everett III was growing up in Washington, D.C., where he read voraciously in science fiction—*Astounding Science Fiction* and other magazines. Fifteen years later he was at Princeton, a graduate student in physics, working with a new thesis advisor: that same John Archibald Wheeler, who must continually reappear, Zelig-like, in the history of time travel. Now it is 1955. Everett is uncomfortable with the idea that simply making a measurement must alter the destiny of a physical system. He makes note of a talk at Princeton in which Einstein says he "could not believe that a mouse could bring about drastic changes in the universe simply by looking at it."* He is also hearing all kinds of dissatisfaction with the various *interpretations* of quantum theory. Niels Bohr's, he feels, is "overcautious." It works, but it doesn't answer the hard questions. "We do not believe that the primary purpose of theoretical physics is to construct 'safe' theories."

So *what if,* he asks—encouraged by Wheeler, who is open as always to the weird and paradoxical—what if every measurement is actually a branching? If a quantum state can be either A or B, then neither possibility is privileged: now there are two copies of the universe, each with its own observers. The world really is a garden of forking paths. Rather than one universe, we have an ensemble of many universes. The cat is definitely alive, in one universe. In another, the cat is

* And by the way, why stop with mice? Can't a machine be an observer? "To draw the line at human or animal observers, i.e., to assume that all mechanical apparata obey the usual laws, but that they are somehow not valid for living observers, does violence to the so-called principle of psycho-physical parallelism," he writes.

dead. "From the viewpoint of the theory," he writes, "all elements of a superposition (all 'branches') are 'actual,' none any more 'real' than the rest." Protective quotation marks run rampant. For Everett, the word *real* is thin ice atop a dark pond:

> When one is using a theory, one naturally pretends that the constructs of the theory are "real" or "exist." If the theory is highly successful (i.e. correctly predicts the sense perceptions of the user of the theory) then the confidence in the theory is built up and its constructs tend to be identified with "elements of the real physical world." This is however a purely psychological matter.

Nonetheless, Everett had a theory, and the theory made a claim: everything that can happen does happen, in one universe or another. New universes are created on demand, as it were. When a radioactive particle may or may not decay, the Geiger counter may or may not register a click, the universe forks again. His dissertation itself followed a difficult path. It exists in several versions. One draft went to Copenhagen, where Bohr did not like it at all. Another, shortened and revised with help from Wheeler, became a paper that could be published in *Reviews of Modern Physics*—despite the obvious objections. "Some correspondents," Everett wrote in a postscript, complained that "our experience testifies" that there is no branching, because we only have one reality. "The argument fails when it is shown that the theory itself predicts that our experience will be what it in fact is," he said—namely, that in *our own* little universe we remain unaware of any branching. When Copernicus theorized that the earth moves, critics objected that we feel no such motion, and they were wrong for precisely the same reason.

Then again, a theory that posits an infinity of universes is an insult to Occam's razor: *Do not multiply entities needlessly.*

Everett's paper did not attract much notice at the time, and it was the last he ever published. He did not continue a career in physics. He died at the age of fifty-one, a chain-smoker and an alcoholic. But perhaps only in this universe. Anyway his theory outlives him. It has acquired a name, the many-worlds interpretation of quantum mechanics, an acronym, MWI, and a considerable following. In its extreme form, this interpretation obviates time altogether. "Time does not flow," says the theorist David Deutsch. "Other times are just special cases of other universes." Nowadays, when parallel worlds or infinite universes are pulled into service as metaphor, they come with semiofficial backing. When someone talks about alternate histories, it could be literature or it could be physics. *The path not taken* and *the road not taken* became common English expressions starting in the fifties and sixties—not earlier, despite Robert Frost's most famous poem. Now any hypothetical scenario can be introduced with the familiar phrase, *In a world where . . .* It becomes harder to remember that this is only a figure of speech.

If we have only the one universe—if the universe is all there is—then time murders possibility. It erases the lives we might have had. Borges knew he was engaging in fantasy. Still, when Hugh Everett was a ten-year-old boy, Borges anticipated the many-worlds interpretation with eight precise words: *"El tiempo se bifurca perpetuamente hacia innumerables futuros."*

Time forks perpetually toward innumerable futures.

EIGHT

Eternity

St. Peter speaks modestly, when he saith, a thousand years to God are but as one day; for, to speak like a Philosopher, those continued instances of time which flow into a thousand years make not to Him one moment: what to us is to come, to His Eternity is present.

—Thomas Browne (1642)

WHAT IF THERE were no such thing as time? What then?

Usually, time travel does not incur physical symptoms—discomfort or illness. In that it differs from air travel, which often causes jet lag. The original time travel of Wells did involve some queasiness:

> I'm afraid I cannot convey the peculiar sensations of time travelling. They are exceedingly unpleasant. There is a feeling exactly like one has on a switchback—of a helpless headlong motion! I felt the same horrible anticipation, too, of an imminent smash.

This is echoed here and there in the literature. Maybe we don't want a magic so profound and consequential to come free of bodily stress.

Ursula K. Le Guin goes a step further in "Another Story; or, A Fisherman of the Inland Sea." Here the travelers obey the laws of physics as we Newtonians and Einsteinians know them. Their spaceships go Nearly as Fast as Light. A journey of four light-years takes just over four years. Relative to the people left behind, the travelers age scarcely at all. If they make an immediate round-trip, on returning home they will seem to have leapt eight years into the future. And how does that feel?

"Of the journey itself," writes Hideo after his first experience, "I have no memory whatever. I think I remember entering the ship, yet no details come to mind, visual or kinetic; I cannot recollect being on the ship. My memory of leaving it is only of an overwhelming physical sensation, dizziness. I staggered and felt sick."

But Hideo's second trip is different. On his second trip, he has the more "usual" experience. It is as if time stops—as if *there is no time.* The journey is a moment—a period? an interval?—in which time does not exist:

> . . . an unnerving interlude in which one cannot think consecutively, read a clockface, or follow a story. Speech and movement become difficult or impossible. Other people appear as unreal half presences, inexplicably there or not there. I did not hallucinate, but everything seemed hallucination. It is like a high fever—confusing, miserably boring, seeming endless, yet very difficult to recall once it is over, as if it were an episode outside one's life, encapsulated.

We've left scientific realism by the roadside. According to relativity, for the people moving at near light speed, time would feel normal. (If time has a normal feeling.) Le Guin is reaching for something else, something unimaginable, the absence of time. When Richard Feynman met a group of schoolchildren and one of them asked what time is, he answered with another question: What if there were no such thing as time? What then?

God knows. He is outside of time, supposedly. He is eternal.

A MAN STEPS INTO a time machine, no need for preliminaries anymore. It has rods, controls, and a starting lever. This one is called "the kettle," and it doesn't resemble a bicycle so much as an elevator. He senses a shimmering, an "unseeable haze," "gray blankness which was solid to the touch, though nonetheless immaterial." He feels a touch of nausea, "the little stir in his stomach, the faint (psychosomatic?) touch of dizziness." The kettle rides in a vertical shaft. So is he going up? Of course not. "Neither up nor down, left nor right, forth nor back." He is going *upwhen*.

By the way—a man, again? Never a woman? Rule: Time travelers are rooted in their authors' time. When our current hero, a Technician named Andrew Harlan, gets into the kettle, he thinks he's a native of the ninety-fifth century, but we recognize him as a man of the year 1955, when Isaac Asimov published his twelfth novel (of forty), *The End of Eternity*. Reading the book now, we can infer some facts of the year 1955:

- Notwithstanding the legacy of H. G. Wells and three decades of pulp magazines, time travel remains a rare and unfamiliar concept to readers in the mainstream. (The *New York Times* went awry by titling its book review "In the Realm of the Spaceman." Spaceman was a better-known concept. The reviewer, Villiers Gerson, raised what he thought was an original question: "If a time traveler were able to go back to 1915 and cause Adolf Hitler to be killed by a bullet in World War I, would our present reality change?" He was neither the first nor the last to wonder.)

- A "computer" is a person who calculates. A reckoner, an arithmetician. A machine for mathematical calculation is called a "computing machine"—in this story, a "Computaplex," capable of "a summation of thousands and thousands of variables." For input and output, a Computaplex uses perforated foil.
- Women are for childbearing. Also for sexual temptation.*

Asimov was just a few years into his career as a science-fiction writer. His first novel, *Pebble in the Sky,* appeared in 1950, when he was a junior professor of biochemistry at the Boston University medical school. It begins with a retired Chicago tailor walking innocently along the street, reciting some verse to himself, when, boom, a nuclear accident in a nearby laboratory transports him fifty thousand years into the future, to a time when Earth is an insignificant planet in the Trantorian Galactic Empire. By then—by 1950, that is—Asimov had sold dozens of stories to *Astounding Science Fiction.* He had been reading the pulps since he discovered them as a child in his father's candy store in Brooklyn. His own origins were, to himself, murky. He knew that his name had originally been Исаак Юдович Озимов, but he never knew his birthday.

As a graduate student, bored by the dissertation he was supposed to be preparing, he invented a chemistry paper titled "The Endochronic

* "Harlan had seen many women in his passages through Time, but in Time they were only objects to him, like walls and balls, barrows and harrows, kittens and mittens."

Properties of Resublimated Thiotimoline," complete with charts, graphs, and citations of nonexistent journals.* The paper describes a made-up substance, thiotimoline, derived from the imaginary bark of a fictitious shrub, which has a mind-bending property dubbed "endochronicity": when placed in water, it dissolves *before* its crystals touch the water. The way quantum mechanics was going, this was only mildly preposterous. Asimov explained it by giving the molecule a peculiar geometrical structure in spacetime: while some of its chemical bonds lie in the usual spatial dimensions, one of them projects into the future and another into the past. You can imagine the possibilities for this quirky crystal. Later, Asimov wrote another paper about its micropsychiatric applications.†

He was soon averaging three to four books a year, but apart from the stage setting blast to the future in *Pebble in the Sky,* he had not tried time travel. The idea that led to *The End of Eternity* came in 1953, when he found a set of bound volumes of *Time* magazine in the stacks of the Boston University library and started reading them through—systematically, from 1928 onward. In one of those early volumes, he was startled to see an advertisement featuring a line drawing with the unmistakable mushroom cloud of a nuclear blast, an image much in people's minds in the fifties, but *not* in the twenties and thirties. When he looked again, he realized he was actually looking at a drawing of the Old Faithful geyser, but by then his mind had already

* The *OED* cites Asimov as the coiner of several words, including *robotics,* but *endochronic* is not one of them. It has not yet caught on.

† Silly? Yet in the distant future—2015—Panasonic marketed a camera that it said recorded images "one second prior to and one second after pressing the shutter button."

leapt to the only other possibility: time travel. Suppose the anachronistic mushroom cloud was some sort of message, sent by a desperate time traveler.

In devising his first novel of time travel, Asimov took the genre in a new direction. This is not the usual hero going on an adventure, hurling himself forth to the future or back to the past. It's a whole universe restructured.

The End of Eternity begins as a play on words, because the one thing everyone knows about eternity is that it has no end. Eternity is everlasting. Traditionally, eternity is God. Or God's bailiwick. (At least in the Judeo-Christian and Islamic traditions, where He is not just eternal but also singular, masculine, and uppercase.) "What times existed which were not brought into being by you?" Augustine asked the Lord in his *Confessions.* "In the sublimity of an eternity which is always in the present, you are before all things past and transcend all things future, because they are still to come." We mortals live in time, but God is beyond that. Timelessness is one of His best powers.

Time is a feature of creation, and the creator remains apart from it, transcendent over it. Does that mean that all our mortal time and history is, for God, a mere instant—complete and entire? For God outside of time, God in eternity, time does not *pass;* events do not occur step by step; cause and effect are meaningless. He is not one-thing-after-another, but all-at-once. His "now" encompasses all time. Creation is a tapestry, or an Einsteinian block universe. Either way, one might believe that God sees it entire. For Him, the story does not have a beginning, middle, and end.

But if you believe in an interventionist god, what does that leave for him to do? A changeless being is hard for us mortals to imagine. Does he *act*? Does he even *think*? Without sequential time, thought—a process—is hard to imagine. Consciousness requires time, it seems. It requires *being in time.* When we think, we seem to think consecutively, one thought leading to another, in timely fashion, forming *memories* all the while. A god outside of time would not have memories. Omniscience doesn't require them.

Perhaps instead an immortal deity is with us in time, enjoying experience, working his will. He sends plagues upon Pharaoh and great winds into the sea, and when the need arises, he sends angels or hornets. Jews and Christians say, "It came to pass in process of time, that the king of Egypt died: and the children of Israel sighed by reason of the bondage. . . . And God heard their groaning, and God remembered His covenant with Abraham." Some theologians would say that when Augustine was confessing, God was listening, and now He remembers. They would say that past is past, for us and for God. If God interacts with our world, it could be in a way that respects our memories of the past and expectations for the future. Perhaps when we discovered time travel, He was suitably amused.

These are deep waters. Even within the Abrahamic religions, theologians have found many divergent ways to speak of God's time or timelessness. All religions, one way and another, conceive of entities whose relation to time transcends our own. "There are two forms of Brahman, time and the timeless," says one Upanishad, though Buddhism is more comfortable than most with the idea that permanence is an illusion:

> Time consumes all beings
> including oneself;
> the being who consumes time,
> cooks the cooker of beings.

The word *eternity* goes back to the beginning, as far as anyone can tell, of our species memory, the beginning of written language. *Aeternus,* in Latin; the Greeks wrote αἰών, which also became *eon.* People needed a word for permanence, or endlessness. Sometimes these words seem to have denoted a duration without beginning or end, or perhaps just without a *known* beginning or end.

No wonder modern philosophers, adapting to a scientific world, continue to torment themselves with such questions. The intricacies multiply. Maybe eternity is like a different reference frame, in the sense made popular by relativity. We have our present moment, and God has a timescale distinct from ours and, indeed, beyond our imagining. Boethius seemed to say something of the kind in the sixth century: "Our 'now,' as though running time, produces a sempiternity, but the divine 'now,' being quite fixed, not moving itself and enduring, produces eternity." Sempiternity is mere endlessness—duration without end. To get outside of time altogether, you need the real thing. "Eternity isn't a long time," the mythologist Joseph Campbell explained. "Eternity has nothing to do with time. . . . The experience of eternity right here and now is the function of life." Or as it is said in Revelations, "There shall be time no longer."

We might decide that the words *outside of time* are a trick of language. Is time a thing to get "outside of," like a box, or a room, or a country—a place invisible to us mortals? In Corinthians it is written: *For thinges which are sene, are temporall: but thynges whiche are not sene, are eternall.*

This last is roughly the premise of Asimov's *End of Eternity.* On the one hand is all humankind, living in time. On the other hand is a place unseen, called Eternity. With a capital E. Only instead of

God, this version of Eternity belongs to a self-selecting group of men. (Again, no women may join the clubhouse. Women are for child-bearing, and this isn't that sort of place.) These men call themselves Eternals, although they are not eternal at all. Nor, as we learn, are they very wise. They engage in backbiting and office politics. They smoke cigarettes. They die. But they act as gods in one way. They have the power to change the course of history, and they use it, again and again. They are compulsive remodelers.

The Eternals form a closed hierarchical society, meritocratic but authoritarian. They are stratified in castes: Computers, Technicians, Sociologists, Statisticians, et al. New arrivals to Eternity, plucked from ordinary Time when young, are Cubs. If they fail in their training, they end up in Maintenance, wearing dun gray uniforms and handling the importation of food and water from Time (even an Eternal has to eat, apparently) and the disposal of waste. Maintenance men are the untouchables, in other words. And how are we to visualize this place, this domain, this realm existing outside of Time? Drearily, it seems rather like an office building: corridors, floors and ceilings, ramps and anterooms. Offices, decorated to suit the taste of the current occupant. An antiquarian might have a bookshelf. ("'Actual books!' He laughed. 'Pages of cellulose, too?'") Most centuries prefer more innovative technology for information storage: "book-films" or "micro-films," which can be spooled through a handy pocket viewer.

Eternity is divided into sections, each associated with a particular century of human history. To go from one section to another, an Eternal rides the kettle: the arrangement feels like stacked floors in a tall skyscraper. Best not to look too closely at the workings. "The laws of the ordinary universe just don't apply to the kettle shafts!"

Between Time and Eternity is a boundary or barrier—an "immaterial" divider—likewise best not examined too closely: "He paused again at the infinitely thin curtain of non-Space and non-Time which separated him from Eternity in one way and from ordinary Time in another." Eternity seems to adjoin the "real" universe anywhere and everywhere. Anyway, transportation from place to place never seems to be a problem. Is Eternity in the fourth dimension? Asimov doesn't bother with the fourth dimension. That's old news. He does tip his hat to the uncertainty principle of quantum mechanics:

> The barrier that separated Eternity from Time was dark with the darkness of primeval chaos, and its velvety non-light was characteristically specked with the flitting points of light that mirrored submicroscopic imperfections of the fabric that could not be eradicated while the Uncertainty Principle existed.

Like Wells not-quite-describing his time machine, Asimov is using his literary wiles to help readers think they are visualizing something that can't be visualized because, after all, it's nonsensical. "Velvety non-light." An artful dodge.* And nice touch, the uncertainty principle decorating the primeval darkness with specks of light.

Now comes a problem of narrative. People live in Eternity, and they do things, one after another, in order to give the story a plot, and before long the fact of narrative makes it impossible to avoid noticing that they (the Eternals), too, operate *in time.* They remember the

* This passage appeared in the first published version of *The End of Eternity* and disappeared from the book version.

past and they worry about the future, just like everyone else. They don't know what's going to happen next. Whatever it would be like to be truly *outside time,* this magical state does not appear conducive to storytelling. Time passes here, too. "Men's bodies grew older and that was the unavoidable measure of time." They call the years "physioyears" and the hours "physiohours." They tell one another, "See you tomorrow." Even in Eternity, they wear wristwatches. It can't be helped.

Since this Eternity is created not by theologians but by technocrats, it does have a beginning and an end. It begins in the twenty-seventh century, after the development of the necessary machinery ("temporal fields" and whatnot), and ends in the "unplumbable entropy death ahead." In the meantime, what fun they have, playing god! The Sociologists profile societies and suggest "reality changes" to fork their history. The Life Plotters diagram the affected lives. The Computers work out the "psycho-mathematics." The Observers go into Time to get data, and the Technicians do the dirty work—e.g., jam the clutch on a vehicle and start a chain of events that prevents a war. When a Technician goes into action, a new branch of possibility becomes real. Then the old branch never happened. It becomes an alternative remembered only in the archives of Eternity.

They believe they are do-gooders.

> We work to plot out all the details of everywhen [explains Technician Harlan] from the beginning of Eternity to where Earth is empty, and we try to plot out all the infinite possibilities of all the might-have-beens and pick out a might-have-been that is better than what is and decide where in Time we can make a tiny little

> change to twist the is to the might-be and we have a new is and look for a new might-be, forever, and forever.

So, for example, Harlan gets out of his kettle, enters Time, and shifts a container from one shelf to another. (He has found the office supplies, apparently.) As a result, a man overlooks something he needs, gets angry, makes a bad decision, a meeting is canceled, a death is postponed—change ripples outward, and some years later what would have been a busy spaceport has vanished from existence. Mission accomplished. If some people must die so that others might live, so be it. You can't make an omelet without breaking eggs, the Eternals have learned. It isn't easy, being responsible for "the happiness of all the human beings who were or ever would be."

What do they value, these masters of the universe? How do they weigh one possible reality against another? It's not always clear. Nuclear war: bad. Drug addiction: bad. Happiness: good, but how to assess it? The Eternals seem to dislike extremes. One century has an excess of hedonism, and Harlan contemplates an improvement: "a different branch of possibility would become real, a branch in which millions of pleasure-seeking women would find themselves transformed into true, pure-hearted mothers." (Lest we forget: these are men of 1950s America.) Mainly, they find themselves continually tinkering with reality in order to eliminate "nuclear technology"—an antiwar measure that has the side effect of keeping humanity from developing interstellar space travel. The reader might guess that the real master of this universe—Isaac Asimov—will vote for space travel.

Without having read Borges, Asimov created a garden of forking paths operated by paper shufflers and bean counters. A branch erased may mean Shakespeare or Bach retroactively unborn, but the Technicians don't care. They pull the plays or the music from Time and store them in the archives.

> Now Harlan stood at the shelves devoted to the novels of Eric Linkollew, usually described as the outstanding writer of the 575th [century], and wondered. He counted fifteen different "Complete Works" collections, each, undoubtedly, taken out of a different Reality. Each was somewhat different, he was sure.

All so futile somehow. The apparatchiks have their own version of Borges's Library of Babel, and it's a storage closet.

With the panorama of history spread out before them, these Eternals have little reason to think about the past. All is future—or is it present? What does it even mean to talk about "the present" in this place? We never really find out. The tinkering with reality just goes on. It is a work in progress.

A few oddballs, though—and our hero, Harlan, is one of them—do take a hobbyist's interest in the centuries before the invention of "temporal fields" and the establishment of Eternity. They call these ancient centuries the Primitive era. No century fascinates them more than the twentieth. Harlan collects Primitive books,

> almost all in print-on-paper. There was a volume by a man called H. G. Wells, another by a man named W. Shakespeare, some tattered histories. Best of all there was a complete set of bound volumes of a Primitive news weekly that took up inordinate space but that he could not, out of sentiment, bear to reduce to micro-film.

Primitive history is locked in place: the Eternals cannot make changes there. "It's like watching history standing still, frozen!" Harlan treasures a verse fragment about a "moving finger," which writes once and then moves on. The Battle of Waterloo has only the one outcome, never to be changed. "That's the beauty of it. No matter what any of us does, it exists precisely as it has always existed." It's so quaint. The technology, too: "In the Primitive era natural petroleum fractions were the source of power and natural rubber cushioned the wheels." Most intriguing—most risible—were the ancients' views of time itself. How could their philosophers be expected to understand? A senior Computer discusses philosophy with Harlan:

> "Now we in Eternity are influenced in our consideration of such things by knowing the facts of Time-travel. Your creatures of the Primitive era, however, knew nothing of Time-travel."
>
> "The Primitives gave virtually no thought to Time-travel, Computer."
>
> "Did not consider it possible, eh?"

Just imagine—people with no concept of time travel! Primitives indeed. The rare exceptions came in the form of "speculations," not by serious thinkers or artists, but only "in some types of escape literature," Harlan explains. "I am not well acquainted with these, but I believe a recurrent theme was that of the man who returned in Time to kill his own grandfather as a child." Yes, that again.

The Eternals know all about the paradoxes. They have a saying: "There are no paradoxes in Time, but only because Time deliberately avoids paradoxes." That grandfather problem arises when you are naïve enough to assume "an indeviant reality" and try to add time travel as an afterthought. "Now your primitives," says the Computer, "never assumed anything *but* an indeviant Reality. Am I right?"

Harlan is not so sure. The escapist literature again. "I don't know enough to answer you with certainty, sir. I believe there may have been speculations as to alternate paths of time or planes of existence."

Bah, says the Computer. That's impossible. "No, without actual experience of Time-travel, the philosophic intricacies of Reality would be quite beyond the human mind."

He has a point. But he underestimates us primitives. We have acquired a rich experience of time travel—a century's worth. Time travel opens our eyes.

MAYBE ASIMOV BEGAN writing this tale optimistically, imagining that a fraternity of wise overseers could nudge humanity onto a better path here and there and steer us away from the nuclear peril that was on everyone's mind in the 1950s. Like Wells, he was a rationalist, a reader of history and believer in social progress. He seems to share the satisfaction his hero, the Technician Harlan, feels in "a universe where Reality was something flexible and evanescent, something men such as himself could hold in the palms of their hands and shake into better shape." If so, Asimov couldn't sustain his optimism. The story takes a dark turn. We begin to see these Eternals not just as philistines but as monsters.

There is a woman after all. Much as Wells's Time Traveller had his girl-of-the-future Weena, Harlan finds Noÿs, "the girl of the 482nd." ("It was not that Harlan had never seen a girl in Eternity before. Never was too strong a word. Rarely, yes . . . But a girl such as *this*!") She has glossy hair, "gluteal curves," milky white skin, and some tinkling jewels that draw attention to her "graceful breasts." She has been assigned to Eternity as a sort of temp for secretarial work. Apparently she is not too bright. Harlan finds he has to explain to her some of the simplest concepts of time. She, in turn, manages to educate him about sex, about which he is naïve, being a stereotype himself.

For a while Noÿs serves as a minor plot device, the motive for some jostling and maneuvering among the Eternals. Harlan, besotted, goes rogue and hustles her into the kettle. They zoom off together. "We're going upwhen, Noÿs." "That means the future, doesn't it?" He stashes her in one of the literature's odder love nests, a spare room in an empty corridor of the year 111,394, where he passes the time with a great deal more *explaining*. He has to explain Reality Changes, he has to explain Computers, he has to explain "physiotime" as opposed to real time. She listens eagerly. "I don't think I'll ever understand it all," sighs Noÿs, her eyes sparkling with "frank admiration."

Eventually he explains his intention to take her with him back in time to *before* the creation of Eternity—to the Primitive era, where they will find themselves in a sparsely populated southwestern territory of the United States of America. "A craggy, lonely world brightened by the splendor of an afternoon sun. There was a soft wind with a chilly edge to it and, most of all, silence . . . bare rocks . . . colored into dull rainbows . . . manless and all but lifeless surroundings."

Harlan thinks he is on a mission to protect Eternity: to close a circle, to ensure its creation. He has a surprise coming: Noÿs is on a mission of her own. She is no Weena. She is an operative sent from a future beyond the imagining even of the Eternals—from a time they have not managed to penetrate, the so-called Hidden Centuries.

It's Noÿs's turn to explain. Her people, the people of the Hidden Centuries, see human history whole, and more than that, as a tapestry of combined possibilities. They see alternative realities as if they were real: "A kind of ghostly never-never land where the might-have-beens play with the ifs." As for Harlan's revered Eternals, she points out that these meddlers are nothing more than a bunch of psychopaths.

> *"Psychopaths!"* exploded Harlan.
>
> "Aren't they? You know them. Think!"

Their incessant petty tinkering has ruined everything, according to the wise future people of the Hidden Centuries. They have "bred out the unusual." In forestalling disasters, they have left no room for triumphs that come only from danger and insecurity. In particular, the Eternals have adamantly prevented the development of nuclear weaponry, at the cost of forestalling any possibility of interstellar travel.

So Noÿs is the time traveler on a mission to change history and Harlan her unwitting pawn. She has brought them on a one-way trip to Primitive times in order to effect the reality change to end all reality changes. She will allow humanity to create its first nuclear explosion at the "19.45th" century, and she will forestall the establishment of Eternity.

Happy ending for Technician Harlan, though: although Noÿs is not the ingenue she has pretended to be, she truly loves him. They will live happily ever after, and "have children and grandchildren, and mankind will remain to reach the stars." We are left with just the one puzzle, then: why the superwoman from the Hidden Centuries, having accomplished her mission of placing humanity on a path to interstellar greatness, wants to settle down with the hapless Andrew Harlan.

So much for eternity. It was a sacred concept: a state of grace, outside of time. For a few hundred pages Asimov turns it into a mere *place*—outside of "Time," but equipped with elevator shafts and storerooms, a uniformed support staff, new men arriving by invitation only. That is quite a comedown. For the godless, though, what else is there? Who has this power over time? The devil.

> With us acts are exempt from time, and we
> Can crowd eternity into an hour,
> Or stretch an hour into eternity.

That's Lucifer, per Lord Byron, on good authority. Luke 4:5: "And the devil, taking him up into an high mountain, showed unto him all the kingdoms of the world in a moment of time." Kurt Vonnegut must have remembered this when he created his Tralfamadorians, adorable green aliens who experience reality in four dimensions: "All moments, past, present and future, always have existed, always will exist. The Tralfamadorians can look at all the different moments just that way we can look at a stretch of the Rocky Mountains, for instance." Eternity is not for us. We may aspire to it, we may imagine it, but we cannot have it.

If we're going to speak literally, nothing is *outside of time.* Asimov ends his story by nullifying it. Who has the privilege of changing history? Not the Technicians, only the author. On the last page the entire previous narrative—the people we have met, the stories we have watched unfold—is erased with the stroke of a pen. The rewriters of history are written out.

NINE

Buried Time

> *So in the future, the sister of the past, I may see myself as I sit here now but by reflection from that which then I shall be.*
>
> —James Joyce (1922)

IN ITS ISSUE of November 1936, *Scientific American* transported readers into the future:

> The time is A.D. 8113. The air channels of the radio-newspaper and world television broadcasting systems have been cleared for an important announcement . . . a story of international importance and significance.

(Evidently it seemed plausible that the world's communications channels could be "cleared" on command.)

> The television sight-and-sound receivers in every home throughout the world carry the thread of the story. In the Appalachian Mountains near the eastern coast of the North American continent is a crypt that has been sealed since the year A.D. 1936. Carefully its contents have been guarded since that date, and today is the day of the opening. Prominent men from all over the world assemble at the site to witness the breaking of the seal that will disclose to the

> waiting world the civilization of an ancient and almost forgotten people.

The ancient and almost forgotten people of 1936 America, that is. This puff was headlined "Today—Tomorrow" and written by Thornwell Jacobs, a former minister and advertising man, now president of Oglethorpe University, a Presbyterian college in Atlanta, Georgia. Oglethorpe had been shuttered since the Civil War. Jacobs re-created it in partnership with a suburban land developer. Now he was promoting his idea, "heartily endorsed" by *Scientific American,* for a Crypt of Civilization, to be waterproofed and sealed in the basement of the administration building on his campus. Jacobs was also a teacher: his course in cosmic history was mandatory for Oglethorpe seniors. Not presuming that Oglethorpe University itself would last forever, he proposed that the crypt should be "deeded in trust to the Federal government, its heirs, assigns, and successors." Its contents? A thorough record of the era's "science and civilization." Certain books, especially encyclopedias, and newspapers preserved in a vacuum or inert gas or on microfilm ("preserved in miniature on motion picture film"). Everyday items such as foods and "even our chewing gum." Miniature models of automobiles. And: "There should also be included a complete model of the capitol of the United States, which, within a half-dozen centuries, will probably have disappeared completely."

Time magazine and *Reader's Digest* picked up the story, and Walter Winchell touted it in one of his radio broadcasts, and the crypt was completed at a ceremony in May 1940. Something about "burying" appealed to people. David Sarnoff of the Radio Corporation of America declared, "The world is now engaged in burying our civilization forever and here in this crypt we leave it to you." The United Press reported:

> ATLANTA, Ga., May 25—They buried the twentieth century here today.
>
> Mickey Mouse and a bottle of beer, an encyclopedia and a movie-fan magazine were put to rest along with thousands of other objects depicting life as it is known today.

Buried our civilization? Buried the twentieth century? The century kept going, making new stuff, even after 1940. What Jacobs really buried was a collection of knickknacks. There was a set of Lincoln Logs children's toys, a sheet of aluminum foil, some women's stockings, model trains, an electric toaster, and phonograph records bearing the voices of Franklin Roosevelt, Adolf Hitler, King Edward VIII, and other world leaders. Some items bound to cause puzzlement: "1 distributor head cover"; "1 sample of catlinite"; "1 lady's breast form." All neatly shelved, a stainless-steel door was welded shut, and so it remains, a quiet room in the basement of what is now called Phoebe Hearst Memorial Hall.*

Imagine how excited the world will be when May 28, 8113, finally arrives.†

* So named to preserve the memory of William Randolph Hearst's mother.

† Why 8113? Jacobs performed some numerology. He reckoned that 6,117 years had passed since the first year of recorded history, which he decided was 4241 BC, according to the Egyptian priestly calendar. Setting 1936 as a midpoint, he did the math and got 8113. It is common for time-capsule buriers to imagine themselves at "the midpoint" of history.

MEANWHILE, the event in Georgia was upstaged by another up north. A public relations man at the Westinghouse Electric and Manufacturing Corporation named G. Edward Pendray—a rocket enthusiast and sometime science-fiction writer—trumped the crypt with a swifter and sleeker package for the future, to be plunged into the ground at the 1939 New York World's Fair—the "World of Tomorrow"—in Flushing, Queens. Instead of a whole room, Westinghouse designed a shiny half-ton torpedo, seven feet long, with an inner glass tube and an outer shell of Cupaloy, a special new alloy of rust-proof hardened copper. Pendray first wanted to call this device a "time bomb," but that term had a different meaning.

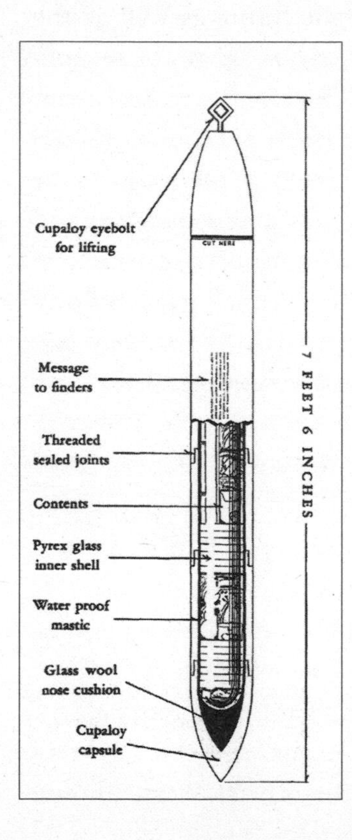

So on second thought he came up with "time capsule." Time, encapsulated. Time in a capsule. A capsule for all time.

The newspapers waxed enthusiastic. "The famous 'time capsule,'" the *New York Times* called it, days after it was announced in the summer of 1938. "Its contents will no doubt prove to be distinctly quaint to the scientists of 6939 A.D.*—as strange, probably, as

* 1939 + 5,000.

the furnishings of Tut-ankh-Amen's tomb seemed to us." The Tutankhamun reference was apt. The burial chamber of the Eighteenth Dynasty pharaoh had been discovered in 1922, causing a sensation: the royal sarcophagus was intact; the British excavators uncovered precious turquoise, alabaster, lapis lazuli, and preserved flowers that disintegrated upon touch. Inner rooms revealed statuettes, chariots, model boats, and wine jars. The pharaoh's funerary mask, solid gold and striped with blue glass, became iconic. So did the very idea of a buried past.

Archeology helped people think about the future as well as the past. Cuneiform tablets were turning up in the desert sands, bearing secrets. The Rosetta Stone, another icon, sat at the British Museum, where for decades no one could read its message—a message to the future, people said, but it hadn't been meant that way. It was for immediate distribution: a decree from king to subjects; pardons and tax rebates. Remember, the ancients had no futurity. They cared less for us than we do for the people of 8113, apparently. Egyptians preserved their treasures and remains for passage to the afterlife, but they weren't waiting for *the future*. They had a different place in mind. Whatever their intent, their eventual legatees were archeologists. So when 1930s Americans began interring their own treasures, they quite self-consciously considered themselves to be enacting archeology in reverse. "We are the first generation equipped to perform our archeological duty to the future," said Thornwell Jacobs.

At the World's Fair, Westinghouse saved space by enclosing 10 million words on microfilm. (They included instructions on how to make a microfilm reader. The time capsule did not have room for one, so a small microscope had to do.) "THE ENVELOPE FOR A MESSAGE TO THE FUTURE BEGINS ITS EPIC JOURNEY," said the official Westinghouse

Book of Record of the Time Capsule of Cupaloy,[*] which was printed and distributed to libraries and monasteries for preservation. Written in an odd faux-biblical prose—as if addressing monks of the Middle Ages, rather than historians of the future—the book advertised the achievements of modern technology:

> Over wires pour cataracts of invisible electric power, tamed and harnessed to light our homes, cook our food, cool and clean our air, operate the machines of our homes & factories, lighten the burdens of our daily labor, reach out and capture the voices and music of the air, & work a major part of all the complex magic of our day.
>
> We have made metals our slaves, and learned to change their characteristics to our needs. We speak to one another along a network of wires and radiations that enmesh the globe, and hear one another thousands of miles away as clearly as though the distance were only a few feet. . . .
>
> All these things, and the secrets of them, and something about the men of genius of our time and earlier days who helped bring them about, will be found in the Time Capsule.

By way of artifacts, the capsule could carry only a few carefully selected items, including a slide rule, a dollar's worth of U.S. coins, and a pack of Camel cigarettes. And one piece of headwear:

[*] Full title: *The Book of Record of the Time Capsule of Cupaloy Deemed Capable of Resisting the Effects of Time for Five Thousand Years; Preserving an Account of Universal Achievements, Embedded in the Grounds of the New York World's Fair, 1939.*

> Believing, as have the people of each age, that our women are the most beautiful, most intelligent, and best groomed of all the ages, we have enclosed in the Time Capsule specimens of modern cosmetics, and one of the singular clothing creations of our time, a woman's hat.

There was also movie footage—or, as the *Book of Record* helpfully explained, "pictures that move and speak, imprisoned on ribbons of cellulose coated with silver."

Several dignitaries were invited to write directly to the people of the future—whoever, whatever, they might be. The dignitaries were grumpy. Thomas Mann informed his distant descendants, "We know now that the idea of the future as a 'better world' was a fallacy of the doctrine of progress." In his message, Albert Einstein chose to characterize twentieth-century humanity this way: "People living in different countries kill each other at irregular time intervals, so that also for this reason anyone who thinks about the future must live in fear and terror." He added hopefully, "I trust that posterity will read these statements with a feeling of proud and justified superiority."

This first time capsule so-called was not the first time anyone thought to hide away some memorabilia, of course. People, like squirrels, are natural hoarders, collectors, and buriers. In the late nineteenth century, amid the rising consciousness of the future, "centennial" fairs inspired time-capsule-like impulses. In 1876 Anna Diehm, a wealthy New York publisher and Civil War widow, set out leather-bound albums for thousands of visitors to sign at the Philadelphia Centennial Exposition and then locked them in an iron safe, along with a gold pen used for the signing and photographs of herself and others, and inscribed a message to posterity: "It is the wish of Mrs. Diehm that this safe may remain closed until July 4, 1976, then to be opened by the Chief Magistrate of the United States."* But the Westinghouse time capsule and the Oglethorpe crypt were the first self-conscious attempts at wholesale cultural preservation for the sake of a notional future—reverse archeology. They mark the beginning of what scholars have called the "golden age" of time capsules: the era when people, worldwide and in increasing numbers, have buried in the earth thousands of parcels, ostensibly intended for the information and education of future creatures unknown. In his study *Time Capsules: A Cultural History,* William E. Jarvis calls them "time-information transfer experiences." They represent a special version of time travel. They also represent a special kind of foolishness.

THE TIME CAPSULE IS a characteristically twentieth-century invention: a tragicomic time machine. It lacks an engine, goes nowhere, sits and waits. It sends our cultural bits and bobs traveling into the future at snail's pace. At our pace, that is. They travel through time in parallel with the rest of us, at our standard velocity of one second per second, one day per day. Only we go about our business of living and decaying, while the time capsules try, ostrichlike, to evade entropy.

Builders of time capsules are projecting something forward into the future, but it's mainly their own imaginations. Like people who buy lottery tickets for the momentary dreams of riches, they get

* Her odd and grandiose wish was granted: she persuaded the Capitol to put the safe in a storeroom under the east steps, and in 1976 the chief magistrate—Gerald R. Ford—was happy to pose for photographers while receiving Mrs. Diehm's offering.

to dream of a time to come when, though long dead, they will be the cynosure of all eyes. "A story of international importance and significance." "Prominent men from all over the world assemble." Clear the airwaves: Dr. Thornwell Jacobs, Oglethorpe University, AD 1936, has something to say.

Looking backward, they misconstrue the intentions of their ancestors. They have the disadvantage of hindsight. Cornerstones of new buildings have long been repositories for inscriptions, coins, and relics, and now, when demolition crews stumble across such items, they mistake them for time capsules and summon journalists and museum curators. For example, in January 2015, many news organizations in the United States and Britain reported the "opening" of what they called "the oldest U.S. time capsule," supposedly left to us by Paul Revere and Sam Adams. This was in fact the cornerstone of the Massachusetts State House, dedicated in 1795 at a ceremony attended by Adams, then the governor, along with Revere and William Scollay, a real-estate developer. The cornerstone memorabilia were wrapped in leather, which naturally deteriorated. In 1855 they were found during foundation repairs and reburied, this time in a brass box the size of a small book, with some extra new coins for good luck, and in 2014 State House workers uncovered the box while trying to trace some water damage. This time, it was thought to be a time capsule. The "air channels of the radio-newspaper and world television broadcasting systems" were not cleared, but several reporters showed up and video cameras rolled as museum conservators examined the contents: five newspapers, a handful of coins, the seal of the Commonwealth of Massachusetts, and a dedicatory plaque. From these items what could be inferred? The Associated Press interpreted them this way:

> Early residents of Boston valued a robust press as much as their history and currency if the contents of a time capsule dating back to the years just after the Revolutionary War are any guide.

"How cool is that?" one of the archivists was quoted as saying. Not very. The correspondent for Boston.com, Luke O'Neil, injected a rare note of skepticism: "*Behold these great wonders from the past!* today's press is proclaiming, *a printed broadsheet newspaper and a currency made out of metal.*" These items had nothing to tell us about Paul Revere or Sam Adams or the life and furniture of post-Revolutionary Boston, nor were they ever meant to. The curators decided to seal them up with plaster once again.

Cornerstone deposits are almost as old as cornerstones. They were not messages to people of the future but votive offerings, a form of magic or sacred ritual. Coins dropped in fountains and wishing wells are votive offerings. Neolithic people entombed axe hoards and clay figurines, Mesopotamians hid amulets in the foundations of Sargon's palace, and early Christians cast tokens and talismans into rivers and buried them in church walls. They believed in magic. So, evidently, do we.

When did eternity, or heaven—the afterlife outside of time—give way to the future? Not all at once. For a while they coexisted. In 1897, the Diamond Jubilee year of Queen Victoria, five plasterers completing the new National Gallery of British Art on the site of the old Millbank Prison penciled a message inside a wall:

> This was placed here on the fourth of June, 1897 Jubilee Year, by the Plasterers working on the Job hoping when this is Found that the Plasterers Association may be still Flourishing. Please let us

Know in the Other World when you get this, so as we can drink your Health.

It was found in 1985 when the Tate Britain (as it had become) did some remodeling. The message remains, preserved on film, in the gallery's archive.

If time capsulists are enacting reverse archeology, they are also engaging in reverse nostalgia. That feeling of sweet longing for past times—with some mental readjustment, we can feel it for our own time, without having to wait. We can create instant vintage automobiles, for example. In 1957, the semicentennial year of Oklahoma statehood, a new Plymouth Belvedere with shiny tail fins was buried in a concrete vault near the statehouse in Tulsa, along with a five-gallon can of gasoline, some Schlitz beer, and some useful trinkets in the glove box. It was to be exhumed fifty years later and awarded to a contest winner. And so it was. But there were better ways to store antique cars. Water seeped in, and what Catherine Johnson, ninety-three years old, and her sister Levada Carney, eighty-eight, received was a rusted shell. Tulsa was undaunted. In 1998 the city laid to rest a Plymouth Prowler, for another fifty years.

The craze has become a business, the "future packaging" industry. Companies offer time capsules in a range of styles, colors, materials, and price points, just as mortuaries market coffins. There are extra charges for engraving and welding. Future Packaging and Preservation promotes Personal Sally, Personal Arnold, Mr. Future, and Mrs. Future cylinders. "Are you on a tight budget? Our *Cylindrical Time Capsule* style may be the most practical choice. Always in stock, these capsules are made of stainless steel, are pre-polished, pre-marked on the bottom with the phrase 'Time Capsule.'" The Smithsonian

Institution offers a list of manufacturers and gives professional tips: argon gas and silica gel are good, PVC and soft solder are bad, and as for electronics, "electronics are a problem." Of course, the Smithsonian has a related business model. Museums conserve and preserve our valuables and our knickknacks for the future. With a difference, of course: museums are alive in the culture. They don't hide the best stuff away underground.

Far more time capsules are buried than are ever recovered. Hermetic as these efforts are, "official" records do not exist, but in 1990 a group of time-capsule aficionados organized a so-called International Time Capsule Society, in hopes of creating a registry. The mailing address and website are at Oglethorpe University. In 1999 they estimated that ten thousand capsules had been buried worldwide and nine thousand of those were already "lost"—but lost to whom? Inevitably the information is anecdotal. The society lists a foundation deposit believed to lie under the Blackpool Tower in Lancashire, England, and says that both "remote sensing equipment" and "a clairvoyant" have failed to find it. The town of Lyndon, Vermont, is supposed to have buried an iron box during its centennial celebration in 1891. A hundred years later, Lyndon officials searched the town vault and other sites, in vain. When the television show *M*A*S*H* ended, its cast members tried to bury some props and costumes in a "time capsule" at the 20th Century Fox parking lot in Hollywood. A construction worker found it almost immediately and tried to give it back to Alan Alda. The time capsulists are trying to use the earth, its basements and graveyards and fens, as a great disorganized filing cabinet, but they have not learned the first law of filing: Most of what is filed never again sees the light of day.

A RESIDENT OF New York City transported a thousand years into the past would not understand a word spoken by the people he encountered. Nor, for that matter, would a resident of London. How can we expect to make ourselves understood to people of the year 6939? Time-capsule creators tend not to worry about linguistic change any more than the science-fiction writers do. But, to their credit, the Westinghouse team did worry about making their time capsule intelligible to the scarcely imaginable recipients of their message. It would be an overstatement to say they solved the problem, but at least they thought about it. They knew that archeologists continued to struggle with ancient Egyptian hieroglyphs a century after the lucky breakthrough provided by the Rosetta Stone. Clay tablets and carved stones still surface bearing scripts from lost languages that defy translation—"proto-Elamite" and "Rongorongo" and others that have not even been named.

So the authors of the *Book of Record of the Time Capsule of Cupaloy* tucked in "A Key to the English Language," by Dr. John P. Harrington, ethnologist, Bureau of Ethnology, Smithsonian Institution, Washington, D.C. It comprised a mouth map (or "Mauth Maep") to help with pronunciation of the "33 sounds of 1938 English," a list of the thousand most common English words, and diagrams to convey elements of grammar.

Also enclosed was an enigmatic one-paragraph story, "The Fable of the Northwind and the Sun," repeated in twenty-five different languages—a little Rosetta Stone to help the archeologists of 6939. An explanatory drawing titled "Tenses" showed a steamship labeled

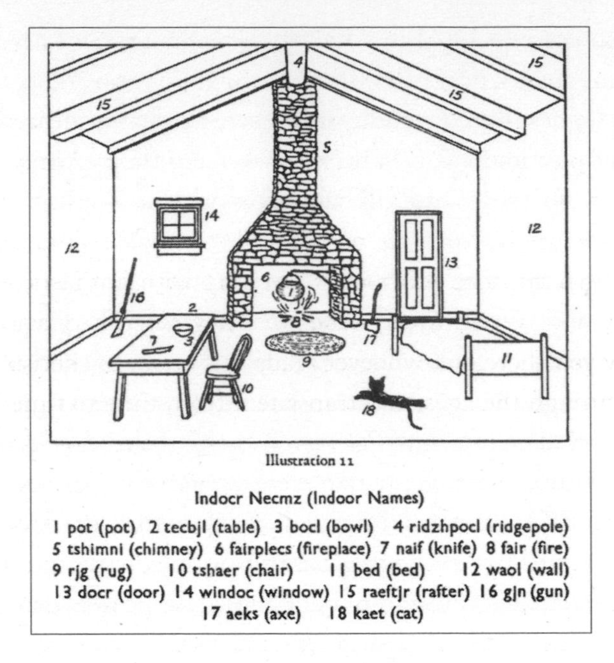

Illustration 11

Indocr Necmz (Indoor Names)

1 pot (pot) 2 tecbjl (table) 3 bocl (bowl) 4 ridzhpocl (ridgepole) 5 tshimni (chimney) 6 fairplecs (fireplace) 7 naif (knife) 8 fair (fire) 9 rjg (rug) 10 tshaer (chair) 11 bed (bed) 12 waol (wall) 13 docr (door) 14 windoc (window) 15 raeftjr (rafter) 16 gjn (gun) 17 aeks (axe) 18 kaet (cat)

present heading from the leftward city (past) to the rightward city (future).

Any effort of this kind confronts a bootstrap problem. The "Key to the English Language" is written, perforce, in English. It uses printed words to explain pronunciation. It specifies sounds in terms of human anatomy. What will our hypothetical future folk make of this: "English has eight vowels (or sounds whose hemming amounts to mere cavity-

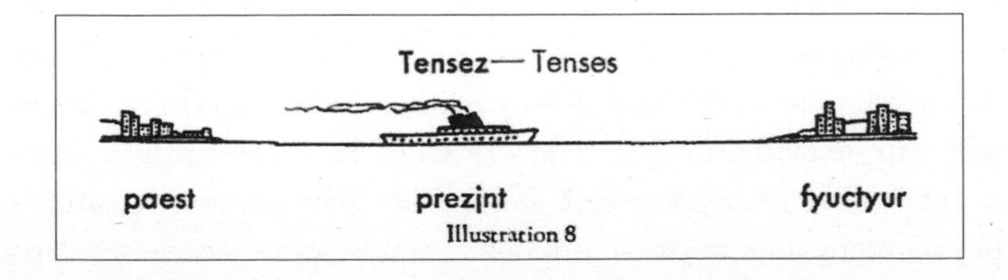

Illustration 8

shape resonance)"? Or this: "The vowel with highest raised back of the tongue, that is, nearest to the *k* consonant position, is *u;* the vowel with the highest raised middle of the tongue, that is, nearest to the *y* consonant position, is *i*"? Who knows where their glottises will be, anyway, or whether those will have gone the way of gills?

The Westinghouse authors also imagined that librarians could continually retranslate the book to keep up with linguistic evolution. And why not? We still read *Beowulf.* They beseeched whomsoever: "We pray you therefore, whoever reads this book, to cherish and preserve it through the ages, and translate it from time to time into new languages that may arise after us, in order that knowledge of the Time Capsule of Cupaloy may be handed down to those for whom it is intended." They would be glad to know that already, as of the twenty-first century, the book is back in print, copyright having been waived: available for around ten dollars from print-on-demand publishers, for ninety-nine cents in an Amazon Kindle version, and widely free online. On the other hand, libraries, short of space, have been "deaccessioning" their copies. Mine once belonged to Columbia University; later it made its way to a used-book dealer in Cleveland, Ohio. Are the librarians forsaking their duty to the future? No, they are fulfilling it, by continually choosing what to keep and what to let go. "We shed as we pick up, like travelers who must carry everything in their arms," says Septimus in Tom Stoppard's *Arcadia,* "and what we let fall will be picked up by those behind. The procession is very long and life is very short. We die on the march. But there is nothing outside the march so nothing can be lost to it. The missing plays of Sophocles will turn up piece by piece, or be written again in another language."

The problem of how to communicate with faraway creatures, physiognomy and language unknown, continues to receive scholarly attention. It arose again when people started sending messages into deep space, in capsules like the Voyager 1 and 2, launched from Cape Canaveral in 1977. These vehicles are space travelers and time travelers, too, their progress measured in light-years. They each bear a copy of the Golden Record, a twelve-inch disk engraved with analog data via the technology, now obsolete, known as "phonograph" (1877–ca. 1987). There are several dozen encoded photographs as well as Sounds of Earth, selected by Carl Sagan and his team and meant to be played at 16⅔ rpm. Just as the Westinghouse time capsule lacked space for a microfilm reader, the Voyager spacecraft could not carry a phonograph record player, but a stylus was thrown in, and the disk is engraved with instructional diagrams. The same conundrum occurs in the context of nuclear-waste disposal: Can we design warning messages to be understood thousands of years hence? Peter C. van Wyck, a communications expert in Canada, described the problem this way: "There is always a kind of tacit assumption that a sign can be made such that it contains instructions for its own interpretation—a film showing how to use a film projector, a map of the mouth to demonstrate pronunciation, recorded instructions for how to assemble and use a stylus and a turntable." If they can figure it all out—decode the information engraved as microscopic waves in a single long spiral groove on a metal disk a half millimeter thick—they will find diagrams of DNA structure and cell division, photographs of anatomy numbered 1–8 from *The World Book Encyclopedia,* human sex organs and a diagram of conception, and an Ansel Adams photograph of the Snake River in Wyoming, and they may "hear" greetings spoken in fifty-five languages (*"shalom"; "bonjour tout le monde"; "namaste"*), sounds of crickets and thunder, a sample of Morse code, and musical selections such as a Bach prelude played by Glenn Gould and a

Bulgarian folk song* sung by Valya Balkanska. That, anyway, is one message sent to deep space and to the far future.

WHEN PEOPLE MAKE time capsules, they disregard a vital fact of human history. Over the millennia—slowly at first and then with gathering speed—we have evolved a collective methodology for saving information about our lives and times and transmitting that information into the future. We call it, for short, culture.

First came songs, clay pots, drawings on cave walls. Then tablets and scrolls, paintings and books. Knots in alpaca threads, recording Incan calendar data and tax receipts. These are external memory, extensions of our biological selves. Mental prostheses. Then came repositories for the preservation of these items: libraries, monasteries, museums; also theater troupes and orchestras. They may consider their mission to be entertainment or spiritual practice or the celebration of beauty, but meanwhile they transmit our symbolic memory across the generations. We can recognize these institutions of culture as distributed storage and retrieval systems. The machinery is unreliable—disorganized and discontinuous, prone to failures and omissions. They use code. They require deciphering. Then again, whether made of stone, paper, or silicon, the technology of culture has a durability that the biological originals can only dream of. This is how we tell our descendants who we were. By contrast, the recent smattering of time capsules is an oddball sideshow.

The capsulists consider it naïve to rely on such perilous and transient human institutions as museums and libraries—all the more so in

* "Izlel je Delyo Hagdutin," or "Delyo the Hajduk Has Gone Outside."

our era of chips and clouds. What good will Wikipedia be when the lights go out, or even the Metropolitan Museum of Art? They believe they are taking the long view. Civilizations rise and fall, with an emphasis on *fall.* From the Bronze Age cultures of the Minoans and the Mycenaeans to the modern civilization in which we live, there was no direct influence—no continuity and no collective memory. These are islands in the ocean of time. So we rely on arrowheads and bones and broken pots found in burial pits. They built their palaces, painted their frescoes, and vanished into obscurity. The darkness drops again. We dig up their remains, but the bits uncovered by archeologists are the accidental bits. In Pompeii it took a cataclysm to freeze vivid tragic tableaus of daily life for our future appreciation. The makers of time capsules prefer not to wait for the sky to pour down ash and pumice.

With the passing millennia, though, humans have developed into something different from the amnesiac creatures who formed those scattered, preliterate settlements. We are well connected information pack rats. Far more mementoes are preserved in museums than in cornerstones. Still more are looked after by coin collectors and random hoarders. The garages of antique automobile collectors are more effective preservers of old cars than buried concrete vaults. Toys? Bottles of old beer? There are specialty museums just for those.

As for knowledge itself, that is our stock in trade. When the Library of Alexandria burned, it was one of a kind. Now there are hundreds of thousands, and they are crammed to overflowing. We have developed a species memory. We leave our marks everywhere. The apocalypse may come—our complacent technocracy foundering amid pandemic or nuclear holocaust or the self-inflicted blighting of the global ecosystem—and when it does, our ruins will be prodigious.

When people fill time capsules they are trying to stop the clock—take stock, freeze the now, arrest the incessant head-over-heels stampede into the future. The past appears fixed, but memory, the fact of it, or the process, is always in motion. That applies to our prosthetic global memory as well as the biological version. When the Library of Congress promises to archive every tweet, does it create a Borgesian paradox in real time or a giant burial chamber in progress?

"BUT IT IS only in ashes that a story endures," wrote the Genovese poet Eugenio Montale. "Nothing persists except extinguished things." When the archeologists of the future come to read our legacy in the proverbial ash heap of history, they will not look to the basement crypt at Oglethorpe University or the time capsule buried in the mud of the former Flushing, Queens. Anyway we will be rewriting that legacy till the bitter end. Stanisław Lem imagined this vividly in his postapocalyptic comic novel *Memoirs Found in a Bathtub,* published in Poland in 1961. The bathtub serves as yet another time capsule. It is marble, "like a sarcophagus," in an intricate complex of corridors (designed by Kafka, evidently) deep underground.* It is buried, more or less apocalyptically, and a millennium or so later it is excavated by archeologists of the future. In it they find a pair of human skeletons and a handwritten manuscript: "a voice speaking to us across the abyss

* "Once again I was walking alone down endless corridors, corridors that continually branched out and converged, corridors with dazzling walls and rows of white, gleaming doors. . . . An endless white labyrinth lay in wait out there, I knew, and an equally endless wandering. The net of corridors, halls and soundproof rooms, each ready to swallow me up . . . the thought made me break out in a cold sweat."

of centuries, a voice belonging to one of the last inhabitants of the lost land of Ammer-Ka."

A faux-scholarly introduction by these future archeologists (or "histognostors") explains the situation. Everyone knows about that turning point in Earth's history called the Great Collapse: "that catastrophic event which in a matter of weeks totally demolished the cultural achievement of centuries." What triggered this Great Collapse was a chemical chain reaction that caused the near-instantaneous disintegration, worldwide, of the peculiar material—"whitish, flaccid, a derivative of cellulose, rolled out on cylinders and cut into rectangular sheets"—called "papyr." Papyr was almost the sole means of recording knowledge: "information of all kinds was impressed on it with a dark tint." Of course nowadays (the histognostors remind their readers) we have metamnestics and data crystallization, but those modern techniques were unknown to this primitive civilization.

> True, there were the beginnings of artificial memory; but these were large, bulky machines, troublesome to operate and maintain, and used only in the most limited, narrow way. They were called "electronic brains," an exaggeration comprehensible only in the historical perspective.

The world's economic systems depended utterly on papyr for regulation and control. Education, work, travel, and finance—all were thrown into disarray when the papyr turned ash. "Panic hit the cities; people, deprived of their identity, lost their reason." After the Great Collapse came the long, dark epoch called the Chaotic. Wandering hordes abandoned the cities. Construction halted (no blueprints). Illiteracy and superstition became universal. "The more complex a civilization," the archeologists note, "the more vital to its existence is the maintenance of the flow of information; hence the more vulnerable it becomes to any disturbance in that flow." Now, and for centuries to come, anarchy prevailed.

This far-future cosmic archeological perspective frames the nearer-future narrative, which we are meant to understand was written in the last days of papyr. The narrator himself seems to be a bewildered civilian navigating a paranoid military bureaucracy. We readers, knowing what we know about the sad fate in store for the written word, may smile grimly as clerks stamp index cards "classified," documents tumble from mail chutes, envelopes shoot through pneumatic tubes, dog-eared folders vanish into metal safes, and paper tape snakes from computers. Of course, we recognize our own world, too.

Rambling deeper and deeper into the labyrinth, the narrator stumbles upon a room full of books: "gray, crumbling" books on dusty, sagging shelves. It is the Library. A balding, shuffling, bespectacled, cross-eyed old man seems to be in charge. He presides over a catalogue of green, pink, and white cards "in no apparent order," stuffed into "endless rows of drawers, their labels framed in brass." On one desk the narrator finds an encyclopedia of heavy black volumes, one lying open to "ORIGINAL SIN—the division of the world into Information and Misinformation." The narrator staggers, dizzy in this darkness broken only by a few naked bulbs. He is overwhelmed by the books' mildewy stench: "this heavy, nauseating breath of the moldering centuries." The old librarian keeps offering him dusty volumes: *Basic Cryptology; Automated Self-Immolation;* "Ah, here is *Homo Sapiens As a Corpus Delicti,* a splendid work, splendid . . ." When he finally escapes this paranoid nightmare of a library, he feels as if he has stepped out of a slaughterhouse.

He is aimless and tired. He keeps looking for orders or instructions. They are not forthcoming. "And so my future remained unknown to me," he muses, "almost as if it hadn't been written down in any ledger anywhere." But we know that his terminal bathtub awaits. He is about to become a time capsule.

TEN

Backward

There are no compasses for journeying in time. As far as our sense of direction in this unchartable dimension is concerned, we are like lost travellers in a desert.

—Graham Swift (1983)

IF YOU COULD take one ride in a time machine, which way would you go?

The future or the past? Sally forth or turn back? ("Right then, Rose Tyler, you tell me," says the Doctor. "Where do you want to go? backwards or forwards in time. It's your choice. What's it going to be?") Do you prefer the costumed pageant of history or the techno-marvels to come? It seems there are two kinds of people. Both camps have their optimists as well as their pessimists. Disease is a worry. Time traveling while black or female poses special hazards. Then again, some people see ways to make money at lotteries, stock markets, and racetracks. Some just want to relive past loves. Many back travelers are driven by regret—mistakes made, opportunities lost.

You may wonder about the rules of this game. Is safety guaranteed? Can you take anything with you?* At the very least, presum-

* Opinions vary. James E. Gunn (1958): "You are naked, because you can take nothing with you, just as you can leave nothing behind. Those are the two natural rules of time travel."

ably, you carry your awareness and your memories, if not a change of clothing. Will you be a passive observer or can you change the course of history? If you change history, does that change you, in turn? "History makes you what you are," says an armchair philosopher in Dexter Palmer's 2016 novel, *Version Control.* "And if you traveled back in time you wouldn't get to be you anymore. You would have a different history, and you would become someone else." The rules keep changing, it seems.

Wells, though he later published not one but two histories of the world, had no interest in sending his Time Traveller backward. He plunged forward, then forward again, on to the end of time. But it didn't take long for other writers to see other possibilities. Edith Nesbit, a friend of Wells's, was a forward-looking, free-thinking fellow socialist, but when she got her chance, the past was the place for her. Writing under the gender-free name of E. Nesbit, she was commonly said to be an author of children's books. Generations later, Gore Vidal took issue with that categorization: he said children were the heroes of her books, but don't be fooled, they are not her ideal readers. He compared her to Lewis Carroll: "Like Carroll, she was able to create a world of magic and inverted logic that was entirely her own." He thought she should be more famous.

Wells often visited the household over which she presided with her husband, Hubert Bland. "The quicksilver wife" is how he perceived her, alongside "the more commonplace, argumentative cast-iron husband." He thought Hubert was something of a fraud, not as bright as Edith, unable to support the family (she did that, with her writing), and a "Seducer": "The astonished visitor came to realize that most of the children of the household were not E. Nesbit's but the results of Bland's conquests . . ."* E. Nesbit became one of the first English writers to explore the new possibilities of time travel. She did not bother with science. There is no machinery, only magic. And where Wells looked forward, she looked back.

Her strange tale *The Story of the Amulet,* written in 1906, begins with four children—Cyril, Robert, Anthea, and Jane—moping through a long summer holiday. They have been left alone in London with old Nurse. Father is in Manchuria and mother in Madeira. They are deprived of liberty and primed for adventure.

Their house is in Bloomsbury, "happily situated between a sandpit and a chalk pit," which means they can walk to the British Museum.† In turn-of-the-century London, this was an institution like none other in the world: a treasure house of antiquities from everywhere England had sent its seaborne colonizers and plunderers. It had the Elgin Marbles, named for the Scottish earl who made off with them

* ". . . that the friend and companion who ran the household was the mother of one of these young people, that young Miss so and so, who played Badminton with a preoccupied air was the last capture of Hubert's accomplished sex appeal. All this E. Nesbit not only detested and mitigated and tolerated, but . . . I think found exceedingly interesting." Then again, Wells himself fathered children with various women besides his wife, and may have had an affair with one of Bland's illegitimate daughters. Free love, after all.

† The book is dedicated to the British Museum's leading Egyptologist, Wallis Budge.

from the Acropolis of Athens. It had the only surviving original of *Beowulf*. Visitors could walk into a gallery and examine the Rosetta Stone on a plinth. The museum was a portal to the past, a time gate through which ancient artifacts poked their age-worn surfaces into modernity: a bronze head from Smyrna, mummy cases from Egypt, winged sphinxes of sandstone, drinking vessels looted from Assyrian tombs, and hieroglyphs preserving secrets in a lost language.

If Cyril, Robert, Anthea, and Jane were getting an education in the perplexities of time—past and present jumbled together in odd ways, cultures misunderstanding one another across a gulf of ages—so were England's adults. Besides museums there were shops trading in relics of the past—"curiosities" and "antiquities"—especially on Wardour Street, Monmouth Street, and Old Bond and New Bond Streets. These physical objects, worn or broken by the years, were like bottles containing messages written by our ancestors, to tell us who they were. "Antiquities are Historie defaced, or some remnants of History, which have casually escaped the shipwrack of time," Roger Bacon had said. By 1900, London had surpassed Paris, Rome, Venice, and Amsterdam as the world's center of trade in antiquities. Nesbit's band of children walk past a curiosity shop near Charing Cross and there discover a small red charm, an amulet of shiny stone. It is trying to tell them something. It has magic powers. Before they know it, they're on their way to that other country, the Past.

First, a few scientific-sounding words to help them along:

> "Don't you understand? The thing existed in the Past. If you were in the Past, too, you could find it. It's very difficult to make you understand things. Time and space are only forms of thought."

Of course Nesbit had read *The Time Machine.* Late in the story, her heroes do dart briefly into the future (using the British Museum as a portal). They find a sort of socialist utopia—all clean and happy and safe and orderly, perhaps to a fault—and encounter a child named Wells, "after the great reformer—surely you've heard of *him*? He lived in the dark ages." With that brief exception, their real adventures take them backward into the Past (always reverently capitalized). They find themselves in Egypt, where children wear no clothes to speak of and tools are made of flint, because no one has heard of iron. They go to Babylon and meet the Queen in her palace of gold and silver, with flights of marble steps and beautiful fountains and a throne with embroidered cushions. She takes time out from throwing people in jail to entertain the time travelers with cold drinks. "I'm simply dying to talk to you, and to hear all about your wonderful country and how you got here, and everything, but I have to do justice every morning. Such a bore, isn't it?" Then it's off to another ancient land, Atlantis: "Great continent—disappeared in the sea. You can read about it in Plato." They find blue sea sparkling in sunlight, white-capped waves lapping marble breakwaters, and the people riding around on great hairy mammoths—not as mild looking as the elephants they were accustomed to seeing at London's zoo.

Archeology catalyzed imaginative literature. Nesbit didn't intend to invent a time-travel subgenre, because she couldn't see into the future, but she did just that. Meanwhile, also in 1906, Rudyard Kipling published a book of historical fantasies called *Puck of Pook's Hill,* with swords and treasures and children transported through the years by the magic of storytelling. C. S. Lewis read Nesbit's *Amulet* when he was a boy in Ireland: "It first opened my eyes to antiquity, the 'dark backward and abysm of time.'" The road that started here led fifty years later to *Peabody's Improbable History,* the television cartoon series that began appearing on *The Rocky and Bullwinkle Show.* Mr. Peabody, the time-traveling beagle, and his boy, Sherman, take their WABAC Machine back to the construction of the pyramids at Gaza, and also to visit Cleopatra, King Arthur, the emperor Nero, Christopher Columbus, and Isaac Newton, at the foot of his apple tree. Anachronism is rampant. The pedagogy is joyously imperfect.* Later still came the cult film *Bill and Ted's Excellent Adventure:* history "rewritten by two guys who can't spell." Some time-tourists go to ogle, others to study history.

All these children—Cyril, Robert, Anthea, Jane, and the boy Sherman—want to go back and see the famous names enacting their famous stories. They serve as proxies for our desire to know what really happened. That desire seems to burn more fiercely when it is partially satisfied. The better that technology gets at capturing and representing our experience of the present, the more we suffer from the fog of ignorance that divides us from lost times. Progress in visualization shows us what we're missing. In Nesbit's time, statues and painted portraits were giving way to photographs. There was a magic in the way they froze an instant of time. Later, the dog Mr. Peabody was of course expert in the new medium of television. Nowadays every modern historian and biographer has felt the desire to send a video camera into the past—to Newton's garden or King Arthur's court—if an actual time machine is not available.

"I've always felt a wonder at old photographs," says Simon Morley.

* For example, Mr. Peabody solemnly explains that Isaac Newton had a brother, Figby, who invented a cookie.

He is a sketch artist, working in advertising in New York, and he is the narrator of *Time and Again,* a 1970 novel (illustrated with sketches and vintage photographs) by Jack Finney, a former New York ad man himself.* Simon deeply feels the inaccessibility of the past, once alive, now lost, taunting us with the few objects and images that survive.

> Maybe I don't need to explain; maybe you'll recognize what I mean. I mean the sense of wonder, staring at the strange clothes and vanished backgrounds, at knowing that what you're seeing was once real. That light really did reflect into a lens from these lost faces and objects. That these people were really there once, smiling into a camera. You could have walked into the scene then, touched those people, and spoken to them. You could actually have gone into that strange outmoded old building and seen what now you never can—what was just inside the door.

It's not just photographs. Someone appropriately sensitized, like Simon, can see the fingers of the past pressing through the cracks of all his existence. In a dense old city like New York, the past is in the stones and the bricks. The relic that triggers Simon's time travels will turn out to be a residential building—not just any apartment house, but a famous one, the Dakota: "like a miniature town . . . gables, turrets,

* The reader may recall an entirely different *Time and Again.* There have been at least three. As the time-travel express got going, in the second half of the twentieth century, publishers must have had a panicky realization that they were using up all the possible titles. They run together in the mind: *Time and Again — Time After Time — From Time to Time — Out of Time — A Rebel in Time — Prisoner of Time — The Depths of Time — The Map of Time — The Corridors of Time — The Masks of Time — There Will Be Time — Time's Eye.* At least four novels have been titled *Time After Time.*

pyramids, towers, peaks . . . acres of slanted surfaces shingled in slate, trimmed with age-greened copper, and peppered with uncountable windows, dormer and flush; square, round, and rectangular; big and small; wide, and as narrow as archers' slits." This will be his portal.

The conceit of *Time and Again* is that time travel to the past can be accomplished with no machinery, no magic, but merely a trick of the mind, a bit of self-hypnosis. If the right subject, a sensitive person like Simon, can rid his memory and purge his surroundings of every trace of the past century, he can translate himself by an act of will into, for example, the year 1882. First he must get into the mood: "There are no such things as automobiles. . . . There are no planes, computers, television, no world in which they are possible. 'Nuclear' and 'electronics' appear in no dictionary anywhere on the face of the earth. You have never heard the name Richard Nixon . . . or Eisenhower . . . Adenauer . . . Stalin . . . Franco . . . General Patton."

Simon (and the reader) are also primed with the now-customary Wells-style pseudologic, to counter the commonsense knowledge that time travel is impossible. Once again, everything we think we know about time is wrong. Here, in 1970, the patter is updated to stand on the authority of Einstein. "How much do you know about Albert Einstein," says Dr. E. E. Danziger, project director, in the role of learned gentleman. "The list of Einstein's discoveries is a considerable one. But I'll skip to this: Presently he said that our ideas about time are largely mistaken." He explains:

> "We're mistaken in our conception of what the past, present and future really are. We think the past is gone, the future hasn't yet

> happened, and that only the present exists. Because the present is all we can see."
>
> "Well, if you pinned me down [says Simon], I'd have to admit that that's how it seems to me."
>
> He smiled. "Of course. To me, too. It's only natural. As Einstein himself pointed out. He said we're like people in a boat without oars drifting along a winding river. Around us we see only the present. We can't see the past, back in the bends and curves behind us. But it's there."
>
> "Did he mean that literally, though? Or did he mean—"

Good question. Did he mean it literally, or was he merely creating an effective mathematical model? No matter. We're moving quickly now, because Danziger has done Einstein one better and invented a way to step out of the boat and walk back.

The reader will discover that what powers this book is the author's raw love of history—for a special time and place, 1880s New York. *Time and Again* has a twisty plot involving blackmail and murder, as well as a time-traveling love triangle, but you sense that what Jack Finney really cared about—drawing it so painstakingly in words and sketches—was the texture of the time: the mortised cut stone that lines Central Park, a gown of wine-red velvet, the *New York Evening Sun* and *Frank Leslie's Illustrated Newspaper,* hitching posts and gas jets and carriage lamps, silk-hatted men and women carrying muffs and wearing button shoes, the astonishing profusion of telegraph wires, in bunches, darkening the downtown sky. "This was the greatest possible adventure," Simon thinks, and you know that Finney thinks so, too.

> I was like a man on a diving board far higher than any other he's ever dared. . . . However cautiously and tentatively, I was about to participate in the life of these times.

Longing for the past resembles the sentiment (or disorder) called nostalgia. Originally, before our newly heightened sense of past and future, nostalgia meant homesickness: "the longing for home which the Physicians have gone so far to esteem a disease under the name of Nostalgia" (Joseph Banks, 1770, per the *Oxford English Dictionary*). Not till the end of the nineteenth century did the word have anything to do with time. But Finney and other writers are not just nostalgic. They are running their fingers through the fabric of history. They are communing with its ghosts. They are reanimating the dead. Long before Finney, Henry James, too, used a redolent old house as a gateway. Just past the turn of the century, while his brother William, the psychologist, was so fascinated with Proust and Bergson, Henry was struggling with a novel he never managed to finish, published after his death as *The Sense of the Past:* a young, fatherless historian, an inherited London house ("a piece of suggestive concrete antiquity"), and a door. There is something special about James's hero, Ralph Pendrel. He is a "victim of the sense of the Past."

"I've been ridden all my life," he says, "by the desire to cultivate some better sense of the past than has mostly seemed sufficient even for those people who have gone in most for cultivating it." He pauses at the fateful door, James tells us—

> perhaps with the supreme pause of the determined diver about to plunge just marked in him before the closing of the door again

placed him on the right side and the whole world as he had known it on the wrong.

Ralph finds himself in another of those bicentury love triangles, fiancée in the present and a fresher, somehow more innocent woman of the past. He is not called a time traveler—not in 1917—but now we know that's what he is.

Old houses were good for the kind of inspiration that sends a person mysteriously into other times. They have attics and basements, where relics lie untouched for ages. They have doors, and when a door opens, who knows what lies beyond? T. S. Eliot, who particularly admired *The Sense of the Past,* saw this: "I am the old house / With the noxious smell and the sorrow before morning, / In which all past is present." In Daphne du Maurier's novel *The House on the Strand,* the house alone is not enough. Time travel requires a drug—a potion comprising equal parts mumbo-jumbo and hocus-pocus: "It has to do with DNA, enzyme catalysts, molecular equilibria and the like—above your head, dear boy, I won't elaborate." When she wrote the story, du Maurier had recently moved to a house called Kilmarth, on a hilltop near the coast of Cornwall, and she remained there, mainly alone, till the end of her life. Kilmarth is the house on the strand. In the novel, it is said to rest on fourteenth-century foundations, and the fourteenth century is the destination of her fictional hero, an unhappily married book publisher named Dick Young. His trip through time (nausea, vertigo) lands him in a landscape of scrubby moor and young, harsh soil. He is stunned by the clarity. There are hooded ploughmen, wimpled ladies, robed monks, and knights on horseback, and Dick finds himself embroiled in a bloody adventure: adultery, betrayal, and murder. Not only that, he knows, because he has consulted the *Encyclopaedia Britannica,* that the Black Death is about to arrive. Yet he is never so alive as in the past.

The House on the Strand appeared in 1969, a year before *Time and Again,* and Dick describes the feeling of both books' narrators when he says, "I had walked about that other world with a dreamer's freedom but with a waking man's perception." They are interlopers in history. They can witness, but they struggle to find out whether they can belong, intervene, or alter the timeline of events. "Could time be all-dimensional," Dick muses, "—yesterday, today, tomorrow running concurrently in ceaseless repetition?" Whatever that means. He's a book publisher, not a physicist.

"Might it not be," says W. G. Sebald in *Austerlitz,* "that we also have appointments to keep in the past, in what has gone before and is for the most part extinguished, and must go there in search of places and people who have some connection with us on the far side of time, so to speak?" This Past, into which so many travelers launch themselves, is a misty place, perhaps even more so than the Future. It can seldom be remembered, must be imagined. Yet here in our information-rich present, the past seems more with us than ever. The more vivid it gets, the more real it seems, the greater the craving. Feeding the addiction are Ken Burnsian documentaries, Renaissance faires, Civil War reenactments, history cable channels, and augmented-reality apps. Anything that "brings the past to life." Under the circumstances, time machines might seem surplus to requirements, but the practitioners of time travel show no signs of slowing down—not in fiction or in film. Woody Allen has explored time travel several times—into the future with *Sleeper* (1973) and then, in 2011, with *Midnight in Paris,* he throws the lever to the past.

His hero, Gil Pender, is a blond Californian and the ideal of the backward-looking obsessive. His friends tease him about his nostalgia, his "denial of the painful present," his "obsession with *'les temps perdus.'*" He is writing a novel, and its opening lines both celebrate and mock the very genre that this movie so self-consciously joins:

> "Out of the Past" was the name of the store, and its products consisted of memories. What was prosaic and even vulgar to one generation had been transmuted by the mere passing of years to a status at once magical and also camp.

His time-slipping portal is not a machine or a house but Paris itself, the whole city, its past so exposed, at every street corner and flea market. To 1920 he goes, and there the modernists understand his sense of dislocation. "I'm from a different time—a whole other era—the future," he explains. "I slide through time." The surrealist Man Ray replies, "Exactly correct—you inhabit two worlds—so far I see nothing strange." The film's central joke is slowly revealed, and it is recursive, time slips within time slips. Nostalgia is eternal. If the twenty-first century yearns for the Jazz Age, the Jazz Age craves the Belle Époque—every age mourning the loss of another age. Woody Allen is neither the first nor the last to see it this way. "The present is always going to seem unsatisfying," Gil learns, "because life itself is unsatisfying."

Travel to the past begins as tourism in the extreme. Complications soon arise. The sightseers start tinkering. We barely learn to read history before we want to rewrite it. Here come the paradoxes—cause and effect going around in loops. Even Nesbit's child heroes see this. When they meet Julius Caesar at his tent in Gaul, peering across the

Channel toward Britain, they can't resist trying to talk him out of dispatching his legions: "We want to ask you not to trouble about conquering Britain; it's a poor little place, not worth bothering about." This backfires, naturally. They end up talking him into it, because *you can't change history,* and we have just witnessed the birth of a time-travel joke that will evolve into higher and higher forms. Thus, a full century after Nesbit, Woody Allen's time traveler in *Midnight in Paris* meets the young Luis Buñuel and can't resist trying to inspire the director with his own future movie.

> GIL: Oh, Mr. Buñuel, I had a nice idea for a movie for you.
> BUÑUEL: Yes?
> GIL: Yeah, a group of people attend a very formal dinner party and at the end of dinner when they try to leave the room, they can't.
> BUÑUEL: Why not?
> GIL: They just can't seem to exit the door.
> BUÑUEL: But, but why?
> GIL: And because they're forced to stay together the veneer of civilization quickly fades away and what you're left with is who they really are—animals.
> BUÑUEL: But I don't get it. Why don't they just walk out of the room?

When the future meets the past, the future has a knowledge advantage. Yet the past is not easily swayed. Mind you, we're talking about our imaginations—the imaginations of professional imaginers, especially. "Time," wrote the novelist Ian McEwan early in his career—

"not necessarily as it is, for who knows that, but as thought has constituted it—monomaniacally forbids second chances." The rules of time travel have been written not by scientists but by storytellers.

WHEN THEY DID start trying to change history, so many of them came up with the perfect plan. They tried to kill Hitler. They are still trying, to this day. It's easy to see why. Others have done great evil and caused great suffering (Stalin, Mao . . .), but one man looms above the others with his combination of monstrosity and charisma. "Adolf Hitler. Hitler, Hitler, Hitler," says Stephen Fry, in his time-travel novel, *Making History.* If only Hitler can be unmade. The entire twentieth century gets a do-over. The idea arose even before the United States entered the war: the July 1941 issue of *Weird Tales* featured a story called "I Killed Hitler" by Ralph Milne Farley, pseudonym for a Massachusetts politician and pulp writer, Roger Sherman Hoar. An American painter resents the German dictator for several reasons and goes back in time to wring the neck of ten-year-old Adolf. (Surprise: the result, when he returns to the present, is not what he expected.) By the end of the 1940s, Hitler's death at the hands of time travelers was already a meme. It is taken for granted in "Brooklyn Project," a 1948 story by Philip Klass, publishing under the name William Tenn. The Brooklyn Project is a secret government experiment in time travel. "As you know," an official explains, "one of the fears entertained about travel to the past was that the most innocent-seeming acts would cause cataclysmic changes in the present. You are probably familiar with the fantasy in its most currently popular form—if Hitler had been killed in 1930." Impossible, he explains. Scientists have proven beyond doubt that time is "a rigid affair, past, present, and future, and nothing in it could be altered." He keeps saying so, even as the project's time-traveling "chronar" makes its way into prehistory and he and his listeners fail to notice that they are now slimy bloated creatures waving purple pseudopods.

Stephen Dedalus says memorably in *Ulysses* that history is a nightmare from which he is trying to awake. Is there no escape? *What if* Julius Caesar had not been murdered on the Senate steps, or Pyrrhus killed in Argos? "Time has branded them," thinks Stephen, "and fettered they are lodged in the room of the infinite possibilities they have ousted. But can those have been possible seeing that they never were? Or was that only possible which came to pass? Weave, weaver of the wind."

Can these eager assassins change history or can't they? For a while, every new story seemed to offer a new theory. Alfred Bester, a New York PR man turned sci-fi writer, invented his own special variation of *you can't change history* for his 1958 story, "The Men Who Murdered Mohammed." The unhappy protagonist, Henry Hassel, angered by discovering his wife "in the arms of" another man, goes on an ever-more-murderous rampage through history, armed with his time machine and a .45-caliber pistol, killing parents and grandparents and historical figures near and far, Columbus, Napoleon, Mohammed (everyone *but* Hitler), and nothing seems to work. The wife continues in her merry ways. Why? Another sad time traveler finally explains:

> "My boy, time is entirely subjective. It's a private matter. . . . We each travel into our own past, and no other person's. There is no universal continuum, Henry. There are only billions of individuals, each with his own continuum; and one continuum cannot affect

the other. We're like millions of strands of spaghetti in the same pot. . . . Each of us must travel up and down his strand alone."

From branching paths to spaghetti strands.

In Stephen Fry's variation, the hero is a student historian named Michael Young. (One wonders—why do our imaginative time-travel writers keep naming their characters *Young*?) In this variation, he hopes to change history not by assassinating Hitler but by sterilizing his father: "The historian as God. I know so much about you, Mr. So-Called Hitler, that I can stop you from being born." And then? Will the twentieth century live happily ever after? ("It was insane of course. I knew that. It couldn't possibly work. You can't change the past. You can't redesign the present.") All you can do is ask, What if? The novelist makes the world. Kate Atkinson's 2013 novel, *Life After Life,* changes the rules yet again. The shooting of Hitler comes in the opening scene: our heroine, Ursula Todd (surname death this time) fires her father's old service revolver at the Führer across the table in a Munich café in 1930. Then she dies, and she keeps on dying, again and again, at different ages, in different ways, always starting over and trying to do it right. Her alternative lives are like strands of spaghetti in a pot. "History is all about 'what ifs,'" someone tells her, as if she didn't know. Someone else urges, "We must bear witness . . . we must remember these people when we are safely in the future." Atkinson, the author, said later, "I am in that future now, and I suppose this book is my bearing witness to the past."

One consequence of Hitler's being the favorite victim of time-traveling assassins is that he keeps on coming back to life. Here he is, living in the Amazon jungle, ninety years old, in George Steiner's novel *The Portage to San Cristóbal of A.H.* Or alive and well in Berlin, still führer of the Greater German Reich, having won World War II in Robert Harris's *Fatherland.* Or syphilitic and senile, in *The Man in the High Castle* by Philip K. Dick—Germany has won the war, because in this history it was the young Franklin D. Roosevelt who was assassinated before he could put his strong hand on the tiller of history. The variations on this theme continue to multiply. As a literary genre, these counterfactual narratives are called alternative history in English, or *ucronía, uchronie,* etc., or allohistory. The labels arose only in the mid to late twentieth century, when the genre began to explode, fed by time travel and branching universes, but in 1930 James Thurber was presciently satirizing it in the *New Yorker* magazine, in his story "If Grant Had Been Drinking at Appomattox" (identified as a follow-up to "If Booth Had Missed Lincoln," "If Lee Had Won the Battle of Gettysburg," and "If Napoleon Had Escaped to America"). Professionals ask similar questions these days. The humor bleeds into academic historiography. It's possible to become quite obsessed with historical contingency. In a comprehensive study, *The World Hitler Never Made,* Gavriel D. Rosenfeld analyzed as many of the Nazi variations as he could find to see how many ended up making history "the same or worse without Hitler as opposed to being better."* There are few happy endings, he found. It is often the writers of science fiction or "speculative fiction" who give us, not only the weirdest, but the most rigorously analyzed approaches to the working of history.

It all might have been different. For want of a nail, the kingdom was lost. I coulda been a contender. Regret is the time traveler's energy bar. If only . . . *something.* Every writer nowadays knows about the but-

* Rosenfeld then started a blog, *The Counterfactual History Review,* and embarked on a collection to be titled *If Only We Had Died in Egypt!: What Ifs of Jewish History.*

terfly effect. The slightest flutter might alter the course of great events. A decade before the meteorologist and chaos theorist Edward Lorenz chose the butterfly for illustrative purposes, Ray Bradbury deployed a history-changing butterfly in his 1952 story "A Sound of Thunder." Here the time machine—the Machine, a vague mess of "silver metal" and "roaring light"—carries paying sightseers on Time Safaris back to the era of the dinosaurs. Apart from the addition of oxygen helmets and intercoms, the time travel itself is pure Wells: "The Machine howled. Time was a film run backward. Suns fled and ten million moons fled after them. . . . The Machine slowed; its scream fell to a murmur." The Safari operators, though, try to be careful about leaving everything unchanged, because they worry about history.

> A little error here would multiply in sixty million years, all out of proportion. . . . A dead mouse here makes an insect imbalance there, a population disproportion later, a bad harvest further on, a depression, mass starvation. . . . Perhaps only a soft break, a whisper, a hair, pollen on the air, such a slight, slight change that unless you looked close you wouldn't see it. Who knows?

In the event, a feckless time-tourist steps on a butterfly: "an exquisite thing, a small thing that could upset balances and knock down a line of small dominoes and then big dominoes and then gigantic dominoes, all down the years across Time."

The butterfly effect, though, is a matter of potential only. Not every flutter in the air leaves its mark on the ages. Most fade to nothing, damped by viscosity. That was Asimov's assumption in *The End of Eternity:* that the effects of tampering with history tend to die out as the centuries pass, perturbations extinguished by friction or dissipation.

His Technician confidently explains: "Reality has a tendency to flow back to its original position." But Bradbury was right and Asimov was wrong. If history is a dynamical system, it's surely nonlinear, and the butterfly effect must obtain. At some places, some times, a slight divergence can transform history. There are critical moments—nodal points. These are where you want to place your lever. History—our real history, that is—must be full of such moments or people, if only we could identify them. We imagine that we can. Births and assassinations, military victories and defeats. We focus on individuals, heroes and villains with an outsize influence. Hence the fascination with Hitler. *If you could kill just one person* . . . By and large, though, the creators of these fantasies have been wise enough to mock the hubris they imply. "Can anyone alter fate?" asks Philip K. Dick in *The Man in the High Castle.* "All of us combined . . . or one great figure . . . or someone strategically placed, who happens to be in the right spot. Chance. Accident. And our lives, our world, hanging on it." Surely some people, some events, some decisions matter more than others. Nodal points must exist, just not necessarily where we think.

Stuck as we are in our own time, most of us aren't trying to make history, much less change it. We take the days one at a time, and history happens. Clive James has said that the greatest poets aspire not to change literary history but only to enrich it. One more reason for the special fascination with Hitler is his playing God. "The Führer was different," thinks Kate Atkinson's Ursula Todd, "he was consciously making history for the future. Only a true narcissist could do that." Beware the politician who aspires to make history. Ursula herself lives in her many moments, one timeline after another: "the future as much a mystery as the past."

We cannot escape the alternative realities, the limitless variations.

The *OED* carefully tells us that the word *multiverse* was *"orig. Science Fiction"* but now, alas, *"Physics"*: "the large collection of universes in the many-worlds interpretation of quantum mechanics . . . in which each in turn of the different possible outcomes occurs." At the same time, entirely apart from quantum theory, we have discovered the pleasure and pain of virtual worlds, inside the computer or the matrix, forcing us to contemplate the possibility that we ourselves are characters in someone else's simulated reality. Or our own. Nowadays, when one speaks of "the real world," it is difficult to refrain from using ironic quotation marks. We inhabit virtual worlds as familiarly and as avidly as the real one. In virtual worlds time travel could not be easier.

Follow me down the rabbit hole into the looping tunnels. William Gibson will be our Virgil. He is reading *The Alteration,* a 1976 novel of alternative history by Kingsley Amis, better known for his comic satires of contemporary Britain. In this world Europe has succumbed to authoritarianism, but never mind Hitler—the papacy is in charge. The Reformation never happened, and the Catholic Church holds much of the world in a theocratic grip. Amis is, of course, investigating in a sidelong way his own, all-too-real world, just as Philip Roth is in *The Plot Against America* and Fry in *Making History.* Amis's story opens in the Cathedral Basilica of St. George at Coverley, "the mother church of all England and of the English Empire overseas." In passing we observe bits of perverted art history: Turner has evidently painted the ceiling "in commemoration of the Holy Victory," Blake has decorated one wall with some holy frescoes, and the choir sings Mozart's Second Requiem, "the crown of his middle age." Science has been suppressed. Although it is 1976, there are wagons and oil lamps, but "matters electrical were held in general disesteem." And there are wheels within wheels.

Lacking science, literature in the world of *The Alteration* has failed to generate science fiction, but the novel's young hero enjoys reading in a disreputable genre known as Time Romance, or TR, for short. TR "appealed to a type of mind." It was illegal but impossible to suppress fully. Inside this genre has evolved a subgenre known as Counterfeit World, CW. In this subgenre, books imagine histories that never happened—alternative histories. Now Gibson will explain:

> Amis accomplishes, as it were in the attic of his novel, a sublime hall-of-mirrors effect. In our world, Philip K. Dick wrote *The Man in the High Castle,* in which the Axis triumphed in World War II. Within Dick's book there is another, imaginary book, *The Grasshopper Lies Heavy,* envisioning a world in which the Allies won, though that world clearly isn't ours. In Amis's counterfeit world, someone called Philip K. Dick has written a novel, *The Man in the High Castle,* imagining a non-Catholic world. Which isn't ours.

And isn't *theirs.* It's hard to keep track. Amis's boy hero, in his world without science, is amazed to read of a counterfactual world where "they use electricity . . . they send messages all over the Earth with it" and Mozart died in 1799 and Beethoven wrote twenty symphonies, and another famous book explains that humans evolved from a thing like an ape. "This business of TR and CW strikes me," says Gibson, "as it plays so artfully through the book, as likely the best Jorge Luis Borges story Jorge Luis Borges never wrote."

The shelves continue to fill with counterfeit worlds. The future becomes the present, and so every futuristic fantasy is slated to become alternative history. When the year 1984 arrived, Orwell's particular surveillance state made the transition from TR to CW. Then 2001 came and went without any noticeable space odysseys. The careful futurist learns to avoid specifying dates. Still our literature and our filmmaking keep breeding new pasts, along with all the putative futures. And so do we all, every day, every night, waking and dreaming in the subjunctive, weighing the options, regretting the might-have-beens.

"DUAL TIME-TRACKS, alternate universes," scoffs a skeptical lawyer in Ursula K. Le Guin's 1971 novel *The Lathe of Heaven*. "Do you see a lot of old late-night TV shows?"

Her troubled client is a man named George Orr. (A tip of the hat to George Orwell, whose special year, 1984, had nearly arrived when Le Guin, in her forties, departed from her previous form to write this strange book.*) When aliens appear, they pronounce his name Jor Jor.

He is an ordinary man—an office worker, apparently placid, milquetoast, conventional. But George is a dreamer. When he was sixteen, he dreamed that his aunt Ethel had been killed in a car crash, and when he awoke he realized that his aunt Ethel *had* been killed in a car crash, weeks earlier. His dream changed reality retroactively. He has "effective dreams"—a sci-fi trope invented here. You might say he carries alternate universes within. Who else does that? The author, for one.

This is a lot of responsibility, and George doesn't want it. He has

* In an odd coincidence, Le Guin had gone to high school with Philip K. Dick, as she realized later. "*Nobody* knew Phil Dick," she told *The Paris Review*. "He was the invisible classmate."

no more control over his dreams than you or I—not conscious control, anyway. (He fears he had resented Ethel's sexual advances.) Increasingly desperate, he doses himself with barbiturates and dextroamphetamine in hopes of suppressing his dreams altogether and ends up in the hands of a psychologist—a dreaming specialist—named William Haber. Haber believes in striving and control; he believes in the power of reason and science. He is plasticoated, like his office furniture. He hypnotizes poor George in an effort to guide his effective dreams and remake reality, step by step. The doctor's office decor seems to have improved. Somehow he has become Director of the Institute.

For the rest of the universe, pulled in the wake of George's dreams, progress is not so simple. Just as quantum theorists may have trouble finding sensible pathways through an unconstrained cornucopia of universes, so might the conscientious novelist. Le Guin does not make matters easy for the reader. She draws us no diagrams.* We must drift in her currents and listen carefully. The music changes. The weather changes. Portland is a city of ceaseless rain, "a downpour of warm soup, forever." Portland enjoys clear air and level sunlight. Was there a dream about President John F. Kennedy and an umbrella? Dr. Haber encourages George to focus on his dread of overpopulation—Portland is a crowded metropolis of three million souls. Or Portland's population has fallen to a hundred thousand, since the Plague Years and the Crash. Everyone remembers those: pollutants in the atmosphere "combining to form virulent carcinogens," the first epidemic, "the riots, and the fuck-ins, and the Doomsday Band, and the Vigilantes." Only George and now Dr. Haber remember multiple realities. "They

* She said to an interviewer, Bill Moyers, "The book is full of dreams and visions, and you are never sure which is which."

took care of the overpopulation problem, didn't they?" George says sarcastically. "We really did it." When are we less the masters of our thoughts than when we dream?

He is not a time traveler. He does not travel through time. He changes it: the past and the future, at once. Much later, sci-fi developed terminology for these conventions, or borrowed them from physics: alternative histories may be called "timelines" or, per William Gibson, "stubs." In any one stub, people are bound to think that their history is the only one that happened. It's not so much that Orr's dream brings a new plague; it's that once he has dreamed, the plague had always happened. He begins to appreciate the paradox. "He thought: In *that* life, yesterday, I dreamed an effective dream, which obliterated six billion lives and changed the entire history of humankind for the past quarter century. But in *this* life, which I then created, I did *not* dream an effective dream." There was always a plague. If this sounds like George Orwell's "We've always been at war with Eastasia," that's no accident. Totalitarian governments also purvey alternate histories.*

The Lathe of Heaven is a critique of a certain kind of hubris—one that every willful creature shares in some degree. It is the hubris of politicians and social engineers: champions of progress who believe we can remake the world. "Isn't that man's very purpose on earth—to do things, change things, run things, make a better world?" says Haber, the scientist, when Orr expresses doubts. Change is good: "Nothing remains the same from one moment to the next, you can't step into the same river twice."

George sees it differently. "We're in the world, not against it," he

* Indeed, it is the essence of doublethink. "This demands a continuous alteration of the past." The literal rewriting of history was Winston Smith's day job, remember, in the Minitrue RecDep (Ministry of Truth Records Department).

says. "It doesn't work to try to stand outside things and run them that way. It just doesn't work, it goes against life." Evidently he is a natural Taoist. "There is a way but you have to follow it. The world *is*, no matter how we think it ought to be."

Having solved overpopulation, Haber tries to use George to bring about peace on Earth. What could go wrong? The Alien Invasion. Sirens, crashes, silvery spaceships. The eruption of Mount Hood. Orr dreams an end to racial strife, to "color problems." Now everyone is gray.

A word from Zhuang Zhou: "Those who dream of feasting wake to lamentation."

It seems there is no way out of this mess—no way based on intention or control—but an unexpected source of wisdom appears: the Aliens. They look like big green turtles. They sense a kindred spirit in Jor Jor, as well they might, since he has presumably dreamed them into existence. They speak in riddles:

> We also have been variously disturbed. Concepts cross in mist. Perception is difficult. Volcanoes emit fire. Help is offered: refusably. Snakebite serum is not prescribed for all. Before following directions leading in wrong directions, auxiliary forces may be summoned.

They sound vaguely Taoist themselves. "Self is universe. Please forgive interruption, crossing in mist."

Reality vies with irreality. George doubts his sanity. He doubts his free will. He dreams of deep seas and crossing currents. Is he the dreamer or the dream?

"Il descend, réveillé, l'autre côté du rêve." (Le Guin is quoting Victor Hugo now.) He descends, awakened, the other side of the dream.

The Alien says: "There is time. There are returns. To go is to return."

"THIS ABOUT TIME being only a thingummy of thought is very confusing," said one of E. Nesbit's wise children, having been initiated into time's new mysteries. "If everything happens at the same time—"

> "It CAN'T!" said Anthea stoutly, "The present's the present and the past's the past."
>
> "Not always," said Cyril. "When we were in the Past the present was the future. Now then!" he added triumphantly.
>
> And Anthea could not deny it.

We have to ask these questions, don't we? Is the world we have the only world possible? Could everything have turned out differently? What if you could not only kill Hitler and see what happens, but you could go back again and again, making improvements, tweaking the timeline, like the weatherman Phil (Bill Murray) in one of the greatest of all time-travel movies, reliving Groundhog Day until finally he gets it right.

Is this the best of all possible worlds? If you had a time machine, would you kill Hitler?

ELEVEN

The Paradoxes

This seems to be a paradox. But one must not think ill of the paradox, for the paradox is the passion of thought, and the thinker without the paradox is like the lover without passion: a mediocre fellow.

—Søren Kierkegaard (1844)

PROPOSITION: *Time travel is impossible, because you'd be able to go back and kill your grandfather, in which case you, the murderer, would never have been born;* and so on, and so on.

We've been here before. We are in the domain of logic, which is, let's remember, a country distinct from the domain of reality. Its inhabitants speak a dialect of their own, resembling natural language and often quite understandable, but full of pitfalls. A thing can be *logically possible* yet *empirically impossible.* If the logicians give us permission to build a time machine, we may still not be able to build it.

I doubt that any phenomenon, real or imagined, has inspired more perplexing, convoluted, and ultimately futile philosophical analysis than time travel has. (Some possible contenders, determinism and free will, are bound up anyway in the arguments over time travel.) The disputation was well under way while H. G. Wells was still alive to be bemused by it. In his classic textbook, *An Introduction to Philosophical Analysis,* John Hospers tackles the question: "Is it logically possible to go back in time—say, to 3000 B.C., and help the Egyptians build the pyramids? We must be very careful about this one." The possibility is easy to state—we habitually use the same words to talk about time as we do when talking about space—and it's easy to imagine. "In fact, H. G. Wells did imagine it in *The Time Machine,* and every reader imagines it with him." (Hospers misremembers *The Time Machine:* "a person in 1900 pulls a lever on a machine and suddenly is surrounded by the world of many centuries earlier.") Hospers was a bit of a kook, actually, who achieved the distinction, unusual for a philosopher, of having received one electoral vote for president of the United States.* His textbook, first published in 1953, remained standard through four editions and forty years.

His answer to the rhetorical question is an emphatic no. Time travel à la Wells is not just impossible, it is *logically* impossible. It is a contradiction in terms. In an argument that runs for four dense pages, Hospers proves this by power of reason.

"How can we be in the 20th century A.D. and the 30th century B.C. *at the same time*? Here already is one contradiction. . . . It is *not* logically possible to be in one century of time and in another century of time at the same time." You may pause to wonder (though Hospers doesn't) whether a trap is lurking in that deceptively common expression "at the same time." The present and the past are different times, therefore they are not the same time, nor *at* the same time. Q.E.D.

That was suspiciously easy. The point of the time-travel fantasy, however, is that the lucky time travelers have their own clocks. Their time can keep running forward, while they travel back to a different

* A rebellious elector in Virginia refused to cast his ballot for the vote winners, Richard Nixon and Spiro Agnew, in 1972 and voted instead for John Hospers, on the Libertarian line.

time as recorded by the universe at large. Hospers sees this but resists it. "People can walk backward in space, but what would 'going backward in time' literally mean?" he asks.

> And if you continue to live, what can you do but get one day older every day? Isn't "getting younger every day" a contradiction in terms—unless, of course, it is meant figuratively, as in "My dear, you're getting younger every day," where it is still taken for granted that the person, while *looking* younger every day, is still *getting older* every day?

(He gives no hint of being aware of F. Scott Fitzgerald's short story in which Benjamin Button does precisely that. Born as a seventy-year-old, Benjamin grows younger every day, until infancy and oblivion. Fitzgerald admitted the logical impossibility. The story has many offspring.)

Time is simple for Hospers. If you imagine that one day you are in the twentieth century and the next day your time machine carries you back to ancient Egypt, he retorts, "Isn't there a contradiction here again? For the next day after January 1, 1969, is January 2, 1969. The day after Tuesday is Wednesday (this is analytic—'Wednesday' is defined as the day that follows Tuesday)" and so on. And he has one final argument, the last nail in time travel's logical coffin. The pyramids were built before you were born. You didn't help. You didn't even watch. "This is an unchangeable fact," says Hospers and adds, "You can't change the past. That is the crucial point: the past is what happened, and you can't make what happened not have happened." We're still in a textbook about analytical philosophy, but you can almost hear the author shouting:

> Not all the king's horses or all the king's men could make what *has* happened *not* have happened, for this is a logical impossibility. When you say that it is logically possible for you (literally) to go back to 3000 B.C. and help build the pyramids, you are faced with the question: did you help them build the pyramids or did you not? The first time it happened you did *not:* you weren't there, you weren't yet born, it was all over before you came on the scene.

Admit it: you didn't help build the pyramids. That's a fact, but is it a logical fact? Not every logician finds these syllogisms self-evident. Some things cannot be proved or disproved by logic. The words Hospers deploys are more slippery than he seems to notice, beginning with the word *time.* And in the end, he's openly assuming the thing he's trying to prove. "The whole alleged situation is riddled with contradictions," he concludes. "When we say we can imagine it, we are only uttering the words, but there is nothing in fact even logically possible for the words to describe."

Kurt Gödel begged to differ. He was the century's preeminent logician, the logician whose discoveries made it impossible ever to think of logic in the same way. And he knew his way around a paradox.

Where a logical assertion of Hospers sounds like this—"It is logically impossible to go from January 1 *to any other day* except January 2 of the same year"—Gödel, working from a different playbook, sounded more like this:

> That there exists no one parametric system of three-spaces orthogonal on the x_0-lines follows immediately from the necessary and sufficient condition which a vector field v in a four-space must satisfy,

> if there is to exist a system of three-spaces everywhere orthogonal on the vectors of the field.

He was talking about world lines in Einstein's space-time continuum. This was in 1949. Gödel had published his greatest work eighteen years earlier, when he was a twenty-five-year-old in Vienna: mathematical proof that extinguished once and for all the hope that logic or mathematics might assemble a complete and consistent system of axioms, powerful enough to describe natural arithmetic and either provably true or provably false. Gödel's incompleteness theorems were built on a paradox and leave us with a greater paradox.* We know that complete certainty must always elude us. We know that for certain.

Now Gödel was thinking about time—"that mysterious and self-contradictory being which, on the other hand, seems to form the basis of the world's and our own existence." Having escaped Vienna after the Anschluss by way of the Trans-Siberian Railway, he settled at the Institute for Advanced Study in Princeton, where he and Einstein intensified a friendship that had begun in the early thirties. Their walks together, from Fuld Hall to Olden Farm, witnessed enviously by their colleagues, became legendary. In his last years Einstein told someone that he still went to the Institute mainly *um das Privileg zu haben, mit Gödel zu Fuss nach Hause gehen zu dürfen,* to have the privilege of walking home with Gödel. For Einstein's seventieth birthday, in 1949, his friend presented him with a surprising calculation: that his field equations of general relativity allow for the possibility of "universes"

* Gödel's proof "is more than a monument," said John von Neumann, "it is a landmark which will remain visible far in space and time. . . . The subject of logic has completely changed its nature and possibilities with Gödel's achievement."

in which time is cyclical—or, to put it more precisely, universes in which some world lines loop back upon themselves. These are "closed timelike lines," or as a physicist today would say, closed timelike curves (CTCs). They are circular highways lacking on-ramps or off-ramps. A closed timelike curve loops back on itself and thus defies ordinary notions of cause and effect: events are their own cause. (The universe itself—entire—would be rotating, something for which astronomers have found no evidence, and by Gödel's calculations a CTC would have to be extremely large—billions of light-years—but people seldom mention these details.)*

If the attention paid to CTCs is disproportionate to their importance or plausibility, Stephen Hawking knows why: "Scientists working in this field have to disguise their real interest by using technical terms like 'closed timelike curves' that are code for time travel." And time travel is sexy. Even for a pathologically shy, borderline paranoid Austrian logician. Almost hidden inside the bouquet of computation, Gödel provided a few words of almost-plain English:

> In particular, if *P, Q* are any two points on a world line of matter, and *P* precedes *Q* on this line, there exists a time-like line connecting *P* and *Q* on which *Q* precedes *P;* i.e., it is theoretically possible in these worlds to travel into the past, or otherwise influence the past.

Notice, by the way, how easy it had already become for physicists and mathematicians to speak of alternative universes. "In these worlds . . . ," Gödel writes. The title of his paper, when he published it

* Also, the Gödelian universe does not expand, whereas most cosmologists are pretty sure that ours does.

in *Reviews of Modern Physics*, was "Solutions of Einstein's Field Equations of Gravitation," and a "solution" is nothing less than a possible universe. "All cosmological solutions with non-vanishing density of matter," he writes, meaning *all possible universes that aren't empty*. "In this paper I am proposing a solution" = *Here's a possible universe for you*. But does this possible universe actually exist? Is it the one we're living in?

Gödel liked to think so. Freeman Dyson, then a young physicist at the Institute, told me many years later that Gödel would ask him, "Have they proved my theory yet?" There are physicists today who will tell you that if a universe has been proved not to contradict the laws of physics, then yes, it is real. A priori. Time travel is possible.

That's setting the bar fairly low. Einstein was more cautious. Yes, he acknowledged, "such cosmological solutions of the gravitation equations . . . have been found by Mr. Gödel." But he added mildly, "It will be interesting to weigh whether these are not to be excluded on physical grounds." In other words, don't follow the math out the window.* Einstein's caution did little to diminish the popularity of Gödel's closed timelike curves among fans of time travel—and in their number we must count logicians, philosophers, and physicists. They wasted little time in launching the hypothetical Gödel rocket ships.

"Suppose our Gödelian spacetime traveller decides to visit his own past and talk to his younger self," wrote Larry Dwyer in 1973. He specifies:

* Gödel's biographer Rebecca Goldstein remarked, "As a physicist and a man of common sense, Einstein would have preferred that his field equations excluded such an Alice-in-Wonderland possibility as looping time."

at t_1, *T* talks to his younger self
at t_2, *T* enters his rocket to begin his journey to the past.
Let t_1 = 1950; t_2 = 1974

Not the most original start, but Dwyer is a philosopher writing in *Philosophical Studies: An International Journal for Philosophy in the Analytic Tradition,* a far cry from *Astounding Stories.* Dwyer has done his homework:

> Science fiction contains an abundance of stories where the plot centres around certain individuals who, having operated complex mechanical devices, find themselves transported back to the past.

Besides reading the stories, he is reading the philosophical literature, beginning with Hospers's proof of the impossibility of time travel. He thinks Hospers is just confused. Reichenbach is confused, too (that would be Hans Reichenbach, author of *The Direction of Time*), and so is Čapek (Milič Čapek, "Time in Relativity Theory: Arguments for a Philosophy of Becoming"). Reichenbach argued for the possibility of "self encounters"—the "younger ego" meets the "older ego," for whom "the same occurrence takes place a second time," and though this may appear paradoxical it is not illogical. Dwyer begs to differ: "It is this sort of talk that has given rise to so much confusion in the literature." Čapek is drawing diagrams with "impossible" Gödelian world lines. Likewise Swinburne, Whitrow, Stein, Gorovitz ("Gorovitz's problems, of course, are all of his own making") and indeed Gödel himself, who misconstrues his own theory.

They all make the same error, according to Dwyer. They imagine that a time traveler could change the past. That cannot happen. Dwyer can live with other difficulties created by time travel: backward causation (effects preceding their causes) and entity multiplication (time travelers and time machines crossing paths with their doubles). But not this. "Whatever else time travel may entail," he says, "it does not involve changing the past." Consider old *T,* using his Gödelian spacetime loop to travel back from 1974 to 1950, when he meets young *T.*

> The encounter is of course recorded twice in the mental history of the time traveller; while young *T*'s reaction to his encounter with *T* may be one of fear, scepticism, joy, etc., *T,* for his part, may or may not recall his feelings when, in his youth, he was confronted by a person claiming to be his older self. Now of course it would be self contradictory to say that *T* does something to young *T* which, by his memory, he knows does not happen to him.

Of course.

Why can't *T* go back and kill his grandfather? Because he did not. It's that simple.

EXCEPT—of course—it's never that simple.

Robert Heinlein, having created his multitude of Bob Wilsons in 1939, punching one another before self-explaining the mysteries of time travel, revisited the paradoxical possibilities twenty years later in a story that outdid all its predecessors. It was titled "'—All You Zombies—'" and published in *Fantasy and Science Fiction* after a *Playboy*

editor turned it down because the sex made him queasy (it was 1959).*
The story has a transgender plot element, a bit forward for the era but necessary to accomplish the time-travel equivalent of a quadruple axel: the protagonist is his (/her) own mother, father, son, and daughter. The title is also the punchline: "I know where I came from—but where did all you zombies come from?"

Could anyone top this? In purely numerical terms—sure. In 1973 David Gerrold, who had been a young television writer for the short-lived (and, later, long-lived) *Star Trek,* published a novel, *The Man Who Folded Himself,* featuring a college student named Daniel who receives a Timebelt from a mysterious "Uncle Jim," complete with instructions. Uncle Jim urges him to keep a diary, and a good thing, too, because life quickly gets complicated. We soon struggle to keep track as the cast of characters expands accordionlike to include Don, Diane, Danny, Donna, ultra-Don, and Aunt Jane—all of whom are (as if you didn't know) the same person, on a looping temporal roller coaster.

So many variations on a theme. The paradoxes multiply almost as fast as the time travelers, but when you look closely, they are all the same. There is just one paradox, wearing different costumes to suit the occasion. Sometimes it is called the bootstrap paradox—a tribute to Heinlein, whose Bob Wilson pulled himself by his bootstraps into his own future. Or the ontological paradox, a conundrum of being and becoming, a.k.a. "Who's your daddy?" People and objects (pocket watches, notebooks) exist without origin or cause. Jane of "'—All You Zombies—'" is her own mother and father, begging the ques-

* The Heinlein story inspired a 2014 film, *Predestination,* with Ethan Hawke and Sarah Snook playing instances of the time traveler.

tion of where her genes came from. Or: in 1935 an American stockbroker finds a Wellsian time machine ("polished ivory and gleaming brass") hidden by palm leaves in the Cambodian jungle ("the land of mystery"); he throws the lever and arrives back in 1925, where the machine is polished up and cached in the palm leaves.* That is its life cycle: a ten-year closed timelike curve. "But where did it come from *originally*?" the stockbroker asks a yellow-robed Buddhist. The wise man explains as if to a dunce: "There never was any 'originally.'"†

Some of the cleverest loops involve pure information. "Mr. Buñuel, I had a nice idea for a movie for you." A book on how to build a time machine arrives from the future. See also: predestination paradox. Trying to change what's bound to happen somehow helps make it happen. In *The Terminator* (1984), a cyborg assassin (played with an idiosyncratic Austrian accent by a thirty-seven-year-old bodybuilder, Arnold Schwarzenegger) travels back in time to kill a woman before she can give birth to the man who is destined to lead a future resistance movement; the cyborg's failure leaves detritus that makes its own creation possible; etc.

In a way, of course, the predestination paradox predates time travel by several millennia. Laius, hoping to defy the prophecy of his own murder, leaves baby Oedipus in the wilderness to die. Tragically, his plan backfires. The idea of the self-fulfilling prophecy is ancient, though the term is new, coined by the sociologist Robert Merton in

* Wells might have admired this descriptive flourish: "The general impression which the contrivance gave was that of unreality. The right-angles, at which various bars joined each other, did not seem to be quite ninety degrees. The perspective was distinctly off; for regardless from which side one viewed it, the more distant side always seemed to be the larger."

† Ralph Milne Farley, "The Man Who Met Himself" (1935). Of course the man is in a ten-year loop, too. He uses the time to make money in the stock market.

1948 to describe an all-too-real phenomenon: "a *false* definition of the situation evoking a new behavior which makes the originally false conception come true." (For example, a warning of gasoline shortages causes panic buying that leads to gasoline shortages.) People have always wondered whether they can escape destiny. Only now, in the era of time travel, we ask whether we can change the past.

All the paradoxes are time loops. They all force us to think about causality. Can an *effect* precede its *cause*? Of course not. Obviously. By definition. "A cause is an object followed by another," David Hume kept saying. If a child receives a measles inoculation and then suffers a seizure, the inoculation may or may not have caused the seizure. The one thing everyone knows for sure is that the seizure didn't cause the inoculation.

But we're not very good at understanding causes. The first person on record as trying to analyze cause and effect by power of ratiocination was Aristotle, who created layers of complexity that have caused confusion ever after. He distinguished four distinct types of causes, which can be named (making allowances for the impossibility of transmillennial translation) the efficient, the formal, the material, and the final. Some of these are hard for us to recognize as causes. The efficient cause of a sculpture is the sculptor, but the material cause is the marble. Both are needed before the sculpture can exist. The final cause is the purpose for which it is made—its beauty, let's say. Considered chronologically, final causes seem to come later. What is the cause of an explosion: the dynamite? the spark? the bank robber? the safecracking? This line of thought tends to strike modern people as pettifogging. (On the other hand, some professionals find Aristotle's vocabulary pitiably primitive. They would not want to discuss causal relations without mentioning immanence, transcendence, individuation, adicity, hybrid causes, probabilistic causes, and causal chains.) Either way, we do well to remember that nothing, when we look closely, has a single unambiguous incontrovertible cause.

Would you accept the assertion that the cause of a rock is that same rock an instant earlier?

"All reasonings concerning matters of fact seem to be founded on the relation of *Cause and Effect,*" said Hume, but he discovered that the reasonings were never easy or certain. Is the sun the cause of a rock's warming? Is an insult the cause of a person's anger? Only one thing could be said for sure: "A cause is an object followed by another." If an effect doesn't *necessarily* follow from a cause, was it a cause at all? The arguments echoed down the corridors of philosophy and continue to echo, despite Bertrand Russell's attempt to settle the matter once and for all in 1913 with an appeal to modern science. "Oddly enough, in advanced sciences such as gravitational astronomy, the word 'cause' never occurs," he wrote. Time for philosophers to get with the program. "The reason why physics has ceased to look for causes is that, in fact, there are no such things. The law of causality, I believe, like much that passes muster among philosophers, is a relic of a bygone age, surviving, like the monarchy, only because it is erroneously supposed to do no harm."

Russell had in mind the hyper-Newtonian view of science described a century earlier by Laplace—the Universe Rigid—in which all that exists is locked together in a machinery of physical laws. Laplace spoke of the past as the *cause* of the future, but if the whole machine chugs along in lockstep, why should we imagine any particular gear or lever to be more causal than any other piece? We may consider the horse to be the cause of the carriage's motion, but that is mere prejudice. Like it or not, the horse, too, is fully determined. Russell had noticed that when physicists write down their laws in mathematical language, time has no inherent directionality. "The law makes no difference between past and future," he wrote. "The future 'determines' the past in exactly the same sense in which the past 'determines' the future."

> "But," we are told, "you cannot alter the past, while you can to some extent alter the future." This view seems to me to rest upon just those errors in regard to causation which it has been my object to remove. You cannot make the past other than it was—true. . . . If you already know what the past was, obviously it is useless to wish it different. But also you cannot make the future other than it will be. . . . If you happen to know the future—e.g. in the case of a forthcoming eclipse—it is just as useless to wish it different as to wish the past different.

And yet, Russell notwithstanding, scientists can no more abandon causation than anyone else. Cigarette smoking causes cancer, whether or not any particular cigarette causes any particular cancer. The burning of oil and coal in the air causes climate change. A mutation in a single gene causes phenylketonuria. The collapse of a burned-out star causes a supernova. Hume was right: "All reasonings concerning matters of fact seem to be founded on the relation of Cause and Effect." Sometimes it's all we talk about. The lines of causality are everywhere, some short and some long, some firm and others tenuous, invisible, interwoven, and inescapable. They all run in one direction, from past to future.

Let's say that one day in 1811, in the town of Teplice, northwestern Bohemia, a man named Ludwig inks a note on a stave in his sketchbook. On an evening in 2011, a woman named Rachel blows a horn in Boston Symphony Hall, with a measurable effect: the air in that room vibrates with a predominant wavelength of 444 cycles per second. Who can deny that, at least in part, the note on paper caused the atmospheric vibrations two centuries later? Using the laws of physics, the path of influence from those molecules in Bohemia to the molecules in Boston would be challenging to compute, even given Laplace's mythical "intelligence which could comprehend all the forces." Yet we can see an unbroken causal chain. A chain of information, if not matter.

Russell did not end the conversation when he declared notions of causality to be relics of a bygone age. Not only do philosophers and physicists continue to wrangle over cause and effect, they add new possibilities to the mix. Retrocausation is now a topic: also known as backward causation or retro-chronal causation. Michael Dummett, a distinguished English logician and philosopher (and reader of science fiction), seems to have given this branch its start with his 1954 paper, "Can an Effect Precede Its Cause?" followed ten years later with his less tentative "Bringing About the Past." Among the questions he raised was this. Suppose he hears on the radio that his son's ship has sunk in the Atlantic. He prays to God that his son should be among the survivors. Has he blasphemed by asking God to undo what has been done? Or is this prayer functionally identical to praying in advance for his son's safe passage?

What might inspire modern philosophers, against all precedent and tradition, to consider the possibility that effects might precede causes? The *Stanford Encyclopedia of Philosophy* offers this answer: "Time Travel." Indeed, all the time-travel paradoxes, births and murders alike, stem from retrocausality. Effects undo their causes.

> The first main argument against the causal order being the temporal order is that temporally backwards causation is possible in cases such as *time travel.* It seems metaphysically possible that a time traveler enters a time machine at time t_1, thus causing her to exit the time machine at some earlier time t_0. Indeed, this looks to be nomologically possible, since Gödel has proved that there are solutions to Einstein's field equations that permit looping pathways.

Not that time travel settles the matter. "A variety of incoherencies might be alleged here," the encyclopedia cautions, "including the incoherency of changing what is already fixed (causing the past), of being both able and unable to kill one's own ancestors, or of generating a causal loop." Brave writers are willing to risk an incoherency or two. Philip K. Dick ran the clocks backward (as it were) in *Counter-Clock World,* and so did Martin Amis in *Time's Arrow.*

We do seem to be traveling in circles.

"THE RECENT RENAISSANCE of wormhole physics has led to a very disturbing observation," wrote Matt Visser, a mathematician and cosmologist in New Zealand in 1994 in *Nuclear Physics B* (the forking path of *Nuclear Physics* devoted to "theoretical, phenomenological, and experimental high energy physics, quantum field theory and statistical systems"). Evidently the "renaissance" of wormhole physics was well established, though these supposed tunnels through spacetime remained (and remain) entirely hypothetical. The disturbing observation was this: "If traversable wormholes exist then it appears to be rather easy to transform such wormholes into time machines."

It was not just disturbing. It was *extremely* disturbing: "This extremely disturbing state of affairs has led Hawking to promulgate his chronology protection conjecture."

Hawking is, of course, Stephen Hawking, the Cambridge physicist who by then had become the world's most famous living scientist, in part because of his dramatic decades-long struggle with an inexorably paralyzing motor neuron disease and in part because of his flair for popularizing the knottiest problems of cosmology. No wonder he was attracted to time travel.

"Chronology Protection Conjecture" was the title of a paper he wrote in 1991 for *Physical Review D*. He explained the motivation as follows: "It has been suggested that an advanced civilization might have the technology to warp spacetime so that closed timelike curves would appear, allowing travel into the past." Suggested by whom? An army of science-fiction writers, of course, but Hawking cited the physicist Kip Thorne (yet another Wheeler protégé) of the California Institute of Technology, who had been working with his graduate students on "wormholes and time machines."

At some point the term "sufficiently advanced civilization" became a trope. As in: Even if we humans can't do it, could a sufficiently advanced civilization? This is useful not just for SF writers but for physicists, too. So Thorne and Mike Morris and Ulvi Yurtsever wrote in *Physical Review Letters* in 1988, "We begin by asking whether the laws of physics permit an arbitrarily advanced civilization to construct and maintain wormholes for interstellar travel." Not coincidentally, twenty-six years later, Thorne served as executive producer and science advisor for the 2014 big-budget movie *Interstellar*. "One can imagine an advanced civilization pulling a wormhole out of the quantum foam," they wrote in the 1988 paper, and they included an illustration captioned "Spacetime diagram for conversion of a wormhole into a time machine." They were contemplating wormholes with mouths in motion: a spaceship might enter one mouth and exit another mouth *in the past.* Fittingly, they concluded by posing a paradox, only this time it isn't the grandfather who dies:

> Can an advanced being measure Schrödinger's cat to be alive at an event *P* (thereby "collapsing its wave function" onto a "live" state), then go backward in time via the wormhole and kill the cat (collapse its wave function onto a "dead" state) before it reaches *P*?

They left that question unanswered.

Hawking stepped in. He analyzed the wormhole physics as well as the paradoxes ("all sorts of logical problems, if you were able to change history"). He considered the possibility of evading the paradoxes "by some modification of the concept of free will," but free will is seldom a happy topic for a physicist, and Hawking saw a better approach: what he proposed to call the chronology protection conjecture. A great deal of calculation was required, and when the calculating was done, Hawking was convinced: the *laws of physics* would protect history from the supposed time travelers. Notwithstanding Kurt Gödel, they must forbid the appearance of closed timelike curves. "It seems there is a chronology protection agency," he wrote sci-fi-ishly, "which prevents the appearance of closed timelike curves and so makes the universe safe for historians." And he concluded with a flourish—the kind of thing Hawking could get away with in the *Physical Review.* He had more than a theory. He had "evidence":

> There is also strong experimental evidence in favor of the conjecture from the fact that we have not been invaded by hordes of tourists from the future.

Hawking is one of those physicists who knows that time travel is impossible but also knows it's fun to talk about. He points out that we are all traveling through time, one second at a time. He describes black holes as time machines, reminding us that gravitation slows the passage of time locally. And he often tells the story of the party he threw for time travelers—invitations sent only after the fact: "I sat there a long time, but no one came."

In fact, the chronology protection conjecture had been floating about long before Stephen Hawking gave it a name. Ray Bradbury, for example, stated it in his 1952 story about time-traveling dinosaur hunters: "Time doesn't permit that sort of mess—a man meeting himself. When such occasions threaten, Time steps aside. Like an airplane hitting an air pocket." Notice that time has agency here: time *doesn't permit,* and time *steps aside.* Douglas Adams offered his own version: "Paradoxes are just the scar tissue. Time and space heal themselves up around them and people simply remember a version of events which makes as much sense as they require it to make."

Perhaps that seems a bit magical. Scientists prefer to credit *the laws of physics.* Gödel thought a robust, paradox-free universe was simply a matter of logic. "Time travel is possible, but no person will ever manage to kill his past self," he told a young visitor in 1972.* "The *a priori* is greatly neglected. Logic is very powerful." At some point

* Rudy Rucker, a mathematician and, later, an author of science fiction.

chronological protection became part of the ground rules. It even became a cliché. In her 2008 story "The Region of Unlikeness," Rivka Galchen can take all that old stage business for granted:

> Science fiction writers have arrived at analogous solutions to the grandfather paradox: murderous grandchildren are inevitably stopped by something—faulty pistols, slippery banana peels, their own consciences—before the impossible deed can be carried out.

Region of unlikeness comes from Augustine: "I perceived myself to be far off from Thee, in the region of unlikeness"—*in regione dissimilitudinis.* He is not fully realized. Nor are any of us, bound as we are in time and space. "I beheld the other things below Thee, and I perceived that they neither altogether are, nor altogether are not." God is eternity, remember, and we are not, much to our sorrow.

Galchen's narrator falls into a friendship with two older men, philosophers maybe, scientists, it's all a bit vague. The relationships are not well defined. The narrator feels that she is a bit undefined herself. The men speak in riddles. "Oh, time will tell," one of them says. And: "Time is our tragedy, the substance we have to wade through as we try to move closer to God." They vanish from her life for a while. She watches the obituary pages. An envelope appears mysteriously in her mailbox—diagram, billiard balls, equations. She thinks of an old joke: "Time flies like an arrow and fruit flies like a banana." One thing becomes clear: everyone in this story knows a lot about time travel. A fateful loop—the same paradox as ever—begins to emerge from the shadows. Some rules are explained: that "contrary to popular movies, travel into the past didn't alter the future, or, rather, that the future was already altered, or, rather, that it was all far more complicated than that." Fate seems to be tugging at her, in a gentle way. Can anyone evade destiny? Look what happened to Laius. All she can say is, "Surely our world obeys rules still alien to our imagination."

WE BEGIN AGAIN. A woman stands at the end of a "pier"—the open-air observation platform at Orly Airport (*la grande jetée d'Orly*), overlooking a sea of concrete on which the great metal jetliners rest, pointed like arrows toward the future. The sun is pale in a charcoal sky. We hear shrill jet blasts, a ghostly choir, murmuring voices. The woman almost smiles as the wind ruffles her hair. A child holds on to the railing, watching the planes on a warm Sunday. He sees the woman raise her hands to her face in horror and sees also, out of the corner of his eye, a blur, a falling shape. *Later, he knew he had seen a man die,* the narrator intones. Not long afterward, World War III begins. A nuclear holocaust destroys Paris, and the rest of the world, too.

This is *La jetée,* a 1962 film by Chris Marker—the pen name of Christian François Bouche-Villeneuve, born in 1921, a philosophy student, a resistance fighter with the Maquis, and then a roving journalist and photographer.* He was rarely photographed without a mask and lived to be ninety-one. In the fifties, after he worked with Alain Resnais on his Holocaust documentary *Night and Fog,* Resnais said: "A theory is making the rounds, and not without some grounds, that Marker could be an extra-terrestrial. He looks like a human, but perhaps he could be from the future or another planet." Marker called *La*

* The self-description he ultimately settled on: "filmmaker, photographer, traveler."

jetée a "photo-novel": it is composed of still photographs, fading and dissolving, shifting points of view, to create, as one critic observed, the "illusion of a time-space continuum." We are told that it is the story of a man marked by a memory from his childhood. *The sudden roar, the woman's gesture, the crumpling body, and the cries of the crowd on the jetty blurred by fear.* The memory—and the marking—make him a candidate for time travel.

Now the world is dead and radioactive. Ruined churches, cratered streets. Survivors inhabit the tunnels and catacombs under Chaillot; a few men rule over prisoners in a camp. They despair. Their one hope lies in finding an emissary to send back to the past. *Space was off limits. The only link with the means of survival passed through Time. A loophole in Time and then maybe it would be possible to reach food, medicine, sources of energy.* Camp scientists experiment cruelly on one prisoner after another, driving them to madness or death, until finally they come to the nameless man "whose story we are telling." What makes this man different from the others is his obsession with the past—with a particular *image* from the past. *If they could imagine or dream of another time, perhaps they would be able to reenter it. The camp police spied even on dreams.* The message here is that time travel is for the imaginative: an idea that recurs in the literature, for example in Jack Finney's *Time and Again.* Time travel begins in the mind's eye. Here, in *La jetée,* it's a matter not just of transportation but of survival. *The human mind balked. To wake up in another time meant to be born a second time, as an adult. The shock would be too much.*

He lies in a hammock. A mask, with electrodes, covers his eyes. A large hypodermic needle injects drugs into his veins, while background voices whisper in German. *He suffers. They continue. On the tenth day, images begin to ooze, like confessions. A peacetime morning. A*

peacetime bedroom—a real bedroom. Real children. Real birds. Real cats. Real graves. On the sixteenth day he is on the pier at Orly. Empty.

Sometimes he sees a woman, who might be the one he seeks. She is standing on the pier, or driving a car, smiling. A headless body is carved in ruined stone. These are images from a timeless world. He recovers from his trance, but the experimenters send him back again.

This time, he is close to her, speaks to her. She welcomes him without surprise. They are without plans, without memories. Time builds itself simply around them, their only landmarks the flavor of the moment and the markings on the walls. They explore a natural history museum, filled with animals from other times. For her, he is a man of mystery—vanishing periodically, wearing a curious necklace, dog tags from the war to come. *She calls him her Ghost.* It occurs to him that in his world, his time, she is already dead.

Many people, seeing *La jetée* with no foreknowledge, are not aware that they are seeing a series of still images. Then, twenty minutes into the film, the woman, asleep, her hair askew on her pillow, opens her eyes, looks directly at the viewer, breathes, and blinks. Time shudders, becomes momentarily real again. The frozen images have been timeless—memories, crystalized. Perhaps memory is the time traveler's subject. Marker once said, "I will have spent my life trying to understand the function of remembering, which is not the opposite of forgetting, but its other side." And he liked to quote George Steiner: "It is not the past that rules us—it is the image of the past." *Jetée* is a pun, too: *j'étais,* I was.

The hero (if that's what he is) carries out a mission not of his choosing. His masters send him not only to the past but then to the future, too. Humans have survived, and so, concealing his eyes behind military-style sunglasses, he begs them to do what is necessary to enable their own existence. They must help, he says. They must: their very survival proves it. Here is paradox again: the narrator says, "This sophism was taken for fate in disguise." When he returns to the past, as we know he must—*somewhere inside him, the memory of a twice-lived fragment of time*—his destination is Orly Airport. It is Sunday. He knows that the woman will be at the end of the pier. The wind ruffles her hair. She almost smiles. As he runs toward her, it occurs to him that somewhere, too, holding on to the railing, will be the child he once was. And then: *There was no way to escape Time.* He understands. *On ne s'évadait pas du Temps.* The future has followed him here. Only at the last instant does he realize whose death he had witnessed as a child.

TWELVE

What Is Time?

Why is it so difficult—so degradingly difficult—to bring the notion of Time into mental focus and keep it there for inspection? What an effort, what fumbling, what irritating fatigue!

—Vladimir Nabokov (1969)

PEOPLE KEEP ASKING what time is, as if the right combination of words could slip the lock and let in the light. We want a fortune-cookie definition, a perfect epigram. Time is "the landscape of experience," says Daniel Boorstin. "Time is but memory in the making," says Nabokov. "Time is what happens when nothing else does"—Dick Feynman. "Time is nature's way to keep everything from happening all at once," says Johnny Wheeler or Woody Allen. Martin Heidegger says, "There is no time."*

What is time? *Time* is a word. The word refers to something, or some things, but surprisingly often the conversation goes off track when people forget whether they're arguing about the word or the thing(s). Five hundred years of dictionaries have created the assumption that every word must have a definition, so what is time? "A non-spatial continuum in which events occur in apparently irreversible succession from the past through the present to the future" (*American*

* *Die Zeit ist nicht.* But he adds, *Es gibt Zeit.* Time is given.

Heritage Dictionary of the English Language, fifth edition). A committee of lexicographers labored over those twenty words and must have debated almost every one. Nonspatial? That word is not to be found in this very dictionary, but all right, time is not space. Continuum? Presumably time is a continuum—but is that known for sure? "Apparently irreversible" seems a hedge. You sense they're trying to tell us something they hope we already know. The challenge is not so much to inform us as to offer some discipline and care.

Other authorities offer entirely different constructions. Not one of them is wrong. What is time? "The general term for the experience of duration," according to the *Encyclopaedia Britannica* (many editions). The very first English dictionary, Robert Cawdrey's in 1604, avoided the problem and skipped right from *thwite* ("shave") to *timerous* ("fearefull, abashed"). Samuel Johnson said "the measure of duration." (And duration? "Continuance, length of time.") A 1960 children's book trimmed the definition to a single word: *Time Is When.*[*]

The people who compose definitions for dictionaries try to avoid the circularity that comes when they use the very word they are defining. With time it's unavoidable. The lexicographers of the *OED* throw up their hands. They divide "time" (only the noun, not the interjection[†] or the obscure conjunction) into thirty-five distinct senses and almost a hundred subsenses, including: *a point in time; an extent of time; a specific period of time; time available . . . ; the amount of time taken up by something;* and *time viewed as a medium through which travel into the past or future is hypothesized or imagined to be possible.*

[*] Beth Gleick, *Time Is When* (Chicago: Rand McNally, 1960). The present author's mother.

[†] "Time!"

("Cf. *time travel.*") They are covering all the bases. Perhaps their best effort is sense number ten: "The fundamental quantity of which periods or intervals of existence are conceived as consisting, and which is used to quantify their duration." Even that definition merely postpones the circularity. *Duration, period,* and *interval* are defined in terms of time. The lexicographers know very well what time is, until they try to define it.

Like all words, *time* has boundaries, by which I don't mean hard and impenetrable shells but porous edges. It maps weirdly between languages. A Londoner might say, "He did it fifty times, at the very least," while in Paris, where the word for time is *temps,* fifty times is *cinquante fois.* Meanwhile, when the weather is good, the Parisian says, *C'est beau temps.* A New Yorker thinks the time and the weather are different things.* And that is just the beginning. Many languages use a separate word to ask "What is the time?" as opposed to "What is time?"

In 1880 the United Kingdom enacted a legislative definition of time, the Statutes (Definition of Time) Act. This declared itself to be "an Act to remove doubts as to the meaning of Expressions relative to Time occurring in Acts of Parliament, deeds, and other legal instruments." It was enacted "by the Queen's most Excellent Majesty, by and with the advice and consent of the Lords Spiritual and Temporal [Time Lords!], and Commons." If only these wise men and woman could have solved the problem by fiat. Removing doubts about the

* "By a curious caprice," wrote the astronomer Charles Nordmann in 1924, "the French language, different from others, designates by a single word, the word *temps,* two very different things: the time which goes by and the weather, or state of the atmosphere. This is one of the peculiarities which give to our language its hermetic elegance, its concentrated sobriety, its elliptic charm."

meaning of time is an ambitious goal. Alas, it turns out that they were not dealing with *What is time?* but only *What is the time?* The time in Great Britain, as defined by the act, is Greenwich mean time.*

What is time? At the dawn of the written word, Plato struggled with the question. "A moving image of eternity," he said. He could name the parts of time: "days and nights and months and years." Moreover:

> When we say that what has become is become and what becomes is becoming, and that what will become is about to become and that the nonexistent is nonexistent—all these are inaccurate modes

* Even this attempt at definition proved tricky. A test case came on August 19, 1898, at 8:15 p.m. (Greenwich mean time), when a man named Gordon was nicked by the police in Bristol for riding his bicycle without a lamp. The local law clearly stated that every person riding a bicycle (which fell under the definition of "carriage") shall carry a lamp, so lighted as to afford adequate means of signaling the approach of the bicycle, during the period between one hour after sunset and one hour before sunrise. On the evening in question, sunset in Greenwich had occurred at 7:13 p.m., so Gordon was caught riding lampless a full hour and two minutes after sunset.

This did not sit well with the accused man, because the sun set ten minutes later in Bristol than in Greenwich: 7:23, not 7:13. Nonetheless, the justices of the city of Bristol, relying on the Statutes (Definition of Time) Act, found him guilty. After all, they reasoned, everyone would benefit by having "a readily ascertained time of lighting up."

With the help of his solicitors, Darley & Cumberland, poor Gordon appealed. The question before the Court of Appeals was described as "an astronomical one." The appellate court saw it his way. They ruled that sunset is not a "period of time" but a physical fact. Justice Channell was insistent: "According to the decision of the Justices, as it stands, a man on an unlighted bicycle may be looking at the sun in the heavens, and yet be liable to be convicted of the offence of not having his lamp lighted an hour after sunset."

> of expression. But perhaps this whole subject will be more suitably discussed on some other occasion.

Here Aristotle, too, found himself in difficulties. "To start, then: the following considerations would make one suspect that it either does not exist at all or barely, and in an obscure way. One part of it has been and is not, while the other is going to be and is not yet." The past has gone out of existence, the future has not yet been born, and time is made up of these "things which do not exist." On the other hand, he said—looked at differently—time seems to be a consequence of change, or motion. It is "the measure" of change. *Earlier* and *later, faster* and *slower*—these are words that are "defined by" time. *Fast* is a lot of motion in a little time, *slow* is a little motion in a lot of time. As for time itself: "time is not defined by time."

Later, Augustine, like Plato, contrasted time with eternity. Unlike Plato, he could hardly stop thinking about time. It obsessed him. His way of explaining was to say that he understood time very well, until the moment he tried to explain. Let us reverse Augustine's process: stop trying to explain and instead take stock of what we know. Time is not defined by time—that needn't paralyze us. When we leave aside the search for epigrams and definitions, it turns out we know a great deal.*

* "If you stop, in dealing with such words, with their definition, thinking that to be an intellectual finality, where are you? Stupidly staring at a pretentious sham! 'Deus est Ens, a se, extra et supra omne genus, necessarium, unum, infinite perfectum, simplex, immutabile, immensum, aeternum, intelligens,' etc.,—wherein is such a definition really instructive? It means less than nothing, in its pompous robe of adjectives."—William James

---

WE KNOW THAT time is imperceptible. It is immaterial. We cannot see it, hear it, or touch it. If people say they perceive the passing of time, that's just a figure of speech. They perceive something else—the clock ticking on the mantel, or their own heartbeat, or other manifestations of the many biological rhythms below the level of consciousness—but whatever time is, it lies outside the grasp of our senses. Robert Hooke made this very point to the Royal Society in 1682:

> I would query by what Sense it is we come to be informed of Time; for all the Information we have from the senses are momentary, and only last during the Impressions made by the Object. There is therefore yet wanting a Sense to apprehend Time; for such a Notion we have.*

Yet we experience time in a way that we do not experience space. Close your eyes, and space disappears: you may be anywhere; you may be big or small. Yet time continues. "I am listening not to Time itself but to the blood current coursing through my brain, and thence through the veins of the neck heartward, back to the seat of private throes which have no relation to Time," says Nabokov. Cut off from the world, with no sensory perception, we may still count the time. Indeed, we habitually quantify time (". . . and yet we conceive of it as

* Hooke proceeded to dig himself into a hole. "I say, we shall find a necessity of supposing some other Organ to apprehend the Impression that is made by Time." What organ? "That which we generally call Memory, which Memory I suppose to be as much an Organ as the Eye, Ear or Nose." Where is this organ, then? "Somewhere near the Place where the Nerves from the other Senses meet."

a Quantity," said Hooke). This leads to a plausible definition: *Time is what clocks measure.* But what is a clock? *An instrument for the measurement of time.** The snake swallows its tail again.

Once we conceive of time as a quantity, we can store it up, apparently. We save it, spend it, accumulate it, and bank it. We do all this quite obsessively nowadays, but the notion is at least four hundred years old. Francis Bacon, 1612: "To choose Time, is to save Time." The corollary of saving time is wasting it. Bacon again: "Prolix and florid Harangues . . . and other personal Speeches are great Wasters of Time." No one would have begun thinking about time as a bankable commodity who was not already familiar with money. *Time hath, my lord, a wallet at his back, / Wherein he puts alms for oblivion.* But is time really a commodity? Or is this just another shabby analogy, along with time the river?

We go back and forth between being time's master and its victim. Time is ours to use, and then we are at its mercy. *I wasted time, and now doth time waste me,* says Richard II; *For now hath time made me his numbering clock.* If you say that an activity *wastes* time, implying a substance in finite supply, and then you say that it *fills* time, implying a sort of container, have you contradicted yourself? Are you confused? Are you committing a failure of logic? None of those. On the contrary, you are a clever creature, when it comes to time, and you can keep more than one idea in your head. Language is imperfect; poetry, perfectly imperfect. We can occupy the time and pass the time in the same breath. We can devour time or languish in its slow-chapp'd power.

* Lee Smolin tries to escape the circularity in *Time Reborn* by redefining "clock": "For our purposes, a clock is any device that reads out a sequence of increasing numbers." Then again, a person counting to one hundred is not a clock.

Newton, who invented the idea of mass, knew that time didn't have any, that it's not a substance, yet he said that time "flows." He wrote this in Latin: *tempus fluit.* The Romans said *tempus fugit,* time flees, or at any rate that motto began appearing on English sundials in the Middle Ages. Newton would have seen that. True, the hours speed by and are gone, once we learn to measure them, but how can time flee? It's another figure of speech. And how can time flow, if it has no substance?

Newton took pains to distinguish two kinds of time. We might call them physical time and psychological time, but he lacked those words, so he had to struggle a bit. The first kind he called, with a flurry of adjectives, "time absolute true and mathematical" (*tempus absolutum verum & Mathematicum*). The other was time as conceived by the common people—the *vulgus*—and this he called "relative" and "apparent." True time—mathematical time—he inferred from a technological feature of his world, the consistency of clocks. He and the clockmakers both leaned on Galileo here—it was Galileo who established that a swinging pendulum of a given length divides time into regular pieces. He measured time by using his pulse. Shortly thereafter, doctors began using clocks to time pulses. The ancients looked to the heavens for measuring time: the sun, the stars, the moon—those were reliable. They gave us our days, months, and years. (When Joshua needed more time to smite the Amorites, he asked God to halt the sun and moon in their tracks—"Sun, stand thou still upon Gibeon; and thou, Moon, in the valley of Ajalon." Who among us has not wanted to stop time?) Now machinery takes over the reckoning.

Another circularity creeps in—a chicken-and-egg problem. Time is how we measure motion. Motion is how we measure time. Newton tried to escape that by fiat. He made Absolute Time axiomatic. He needed a reliable backbone for his laws of motion. The first law: an object moves at a constant velocity, unless acted upon by some external force. But what is velocity? Distance per unit time. When Newton declared that time flows equably, *aequabiliter fluit,* he meant that we can count on unit time. Hours, days, months, years: they are the same everywhere and always. In effect, he imagined the universe as its own clock, the cosmic clock, perfect and mathematical. He wanted to say that when two of our earthly clocks differ, it's because of some fault in the clocks, not because the universe speeds up and slows down hither and yon.

NOW IT IS fashionable among physicists and philosophers to ask whether time is even "real"—whether it "exists." The question is debated at conferences and symposia and analyzed in books. I have put quotation marks around those words because they are so problematic in themselves. The nature of reality hasn't been settled either. We know what it means to say that unicorns are not real. Likewise Santa Claus. But when scholars say time is not real, they mean something different. They haven't lost faith in their wristwatches or their calendars. They use "real" as code for something else: absolute, special, or fundamental.

Not everyone would agree that physicists like to debate the reality of time. Sean Carroll writes, "Perhaps surprisingly, physicists are not overly concerned with adjudicating which particular concepts are 'real' or not." Leave that to philosophers, I think he means. "For concepts like 'time,' which are unambiguously part of a useful vocabulary we have for describing the world, talking about 'reality' is just a bit of harmless gassing." The business of physicists is to construct theoretical models and test them against empirical data. The models are effective and powerful but remain artificial. They themselves are a kind of language. Still, physicists do get caught up in debating the nature of reality. How could they not? "The nature of time" was the subject of an international essay contest organized in 2008 by FQXi, an institute devoted to foundational questions of physics and cosmology. One winning essay, chosen from more than a hundred, was Carroll's own: "What If Time Really Exists?" This was a deliberately contrarian exercise. "There is a venerable strain of intellectual history that proclaims that time does not exist," he noted. "There is a strong temptation to throw up one's hands and proclaim the whole thing is an illusion."

A landmark on that road is an essay published in 1908 by the journal *Mind,* "The Unreality of Time," by John McTaggart Ellis McTaggart. He was an English philosopher, by then a fixture at Trinity College, Cambridge.* McTaggart was said (by Norbert Wiener) to have made a cameo appearance in *Alice's Adventures in Wonderland* as the Dormouse, "with his pudgy hands, his sleepy air, and his sidelong walk." He had been arguing for years that our common view of time is an illusion, and now he made his case. "It doubtless seems highly paradoxical to assert that Time is unreal," he began. But consider . . .

He contrasts two different ways of talking about "positions in time" (or "events"). We may talk about them relative to the present—

* McTaggart's name bears explaining. He was christened (by his parents, the Ellises of Wiltshire) John McTaggart Ellis, after his father's uncle, Sir John McTaggart, a childless Scottish baronet. Sir John then bequeathed a considerable fortune to the Ellises on the condition that they take his surname. In the case of young John, this led to a redundancy. The double dose of "McTaggart" never seems to have bothered him, and he, not the baronet, is the McTaggart most remembered today.

the speaker's present. The death of Queen Anne (his example) is in the past, for us, but at one time it lay in the future and then came round to the present. "Each position is either Past, Present, or Future," writes McTaggart. This he labels, for later convenience, the A series.

Alternatively, we may talk about the positions in time relative to one another. "Each position is Earlier than some, and Later than some, of the other positions." The death of Queen Anne is later than the death of the last dinosaur but earlier than the publication of "The Unreality of Time." This is the B series. The B series is fixed. It is permanent. The order can never change. The A series is changeable: "an event, which is now present, was future and will be past."

Many people found this A series and B series distinction persuasive, and it lives on robustly in the philosophical literature. By a chain of reasoning McTaggart uses it to prove that time does not exist. The A series is essential to time, because time depends on change, and only the A series allows for change. On the other hand, the A series contradicts its own premises, because the same events possess the properties of pastness and futureness. "Neither time as a whole, nor the A series and B series, really exist" is his apparently inevitable conclusion. (I could say "was" because the paper appeared in 1908. But I can also say "is" because the paper exists in libraries and online and, more abstractly still, in the fast-expanding tapestry of interwoven ideas and facts that we call our culture.)

You may have noticed—and if so, you're more observant than most of his readers—that McTaggart began by assuming the thing he is trying to prove. He considered all *positions in time,* all possible *events,* as if they were already laid out in a sequence, points on a geometer's line, *M, N, O, P,* arranged from the point of view of God or the logician. Call this the eternal point of view, or eternalism. The future is just like the past: you can see it in the mind's eye, neatly diagrammed. Our experience to the contrary is merely a product of mental states: memories, perceptions, and anticipations, which we experience as "pastness," "presentness," and "futurity." An eternalist says that reality is timeless. So time is unreal.

In fact this is a mainstream view of modern physics. I won't say *the* mainstream view—in these tempestuous days no one can say for sure what that is. Many of the most respected and established physicists espouse the following:

- The equations of physics contain no evidence for a flow of time.
- The laws of science do not distinguish between the past and the future.
- *Therefore*—do we have a syllogism?—
- Time is not real.

The observer—physicist or philosopher—stands outside and looks in. The human experience of time is suspended for abstract observation. Past, present, and future are bounded in a nutshell.

And what of our persistent impressions to the contrary? We experience time in our bones. We remember the past, we await the future. But the physicist notes that we are fallible organisms, easily fooled and not to be trusted. Our prescientific ancestors experienced the flat earth and traveling sun. Could our experience of time be equally naïve? Perhaps—but scientists have to come back to the evidence of our senses in the end. They must test their models against experience.

"People like us, who believe in physics," Einstein said, "know that the distinction between past, present, and future is only a stubbornly persistent illusion." *Who believe in physics*—I detect something wistful in that. "In physics," repeats Freeman Dyson, "the division of space-time into past, present, and future is an illusion." These formulations retain a bit of humility that is sometimes lost in the quoting. Einstein was consoling a bereaved sister and son, and perhaps thinking as well of his own pending mortality. Dyson was expressing hopeful bonds of kinship with people of the past and people of the future: "They are our neighbors in the universe." These are beautiful thoughts, but they were not intended as final statements about the nature of reality. As Einstein himself said on an earlier occasion, "Time and space are modes by which we think, and not conditions in which we live."

There is something perverse about a scientist's believing that the future is already complete—locked down tight, no different from the past. The first motivation for the scientific enterprise, the prime directive, is to gain some control over our headlong tumble into an unknown future. For ancient astronomers to forecast the movements of heavenly bodies was vindication and triumph; to predict an eclipse was to rob it of its terror; medical science has labored for centuries to eradicate diseases and extend the lifetimes that fatalists call fixed; in the first powerful application of Newton's laws to earthly mechanics, students of gunnery computed the parabolic trajectories of cannonballs, the better to send them to their targets; twentieth-century physicists not only managed to change the course of warfare but then dreamt of using their new computing machines to forecast and even control the earth's weather. Because, why not? We are pattern-recognition machines, and the project of science is to formalize our intuitions, do the math, in hopes not just of understanding—a passive, academic pleasure—but of bending nature, to the limited extent possible, to our will.

Remember Laplace's perfect intelligence, vast enough to comprehend all the forces and the positions and to submit them to analysis. "To it nothing would be uncertain, and the future as the past would be present to its eyes." This is how the future becomes indistinguishable from the past. Tom Stoppard joins the parade of philosophers wittily paraphrasing him: "If you could stop every atom in its position and direction, and if your mind could comprehend all the actions thus suspended, then if you were really, *really* good at algebra you could write the formula for all the future; and although nobody can be so clever as to do it, the formula must exist just as if one could." It bears asking—because so many modern physicists still believe something like this—why? If no intelligence can be so comprehensive, no computer can do so much computing, why must we treat the future as though it were predictable?

The implicit answer, sometimes explicit, is that the universe is its own computer. It computes its own destiny, step by step, bit by bit (or qubit by qubit). The computers we know, in the early twenty-first century, not counting the tantalizing quantum variety, operate deterministically. A given input always leads to the same output. Our input, again, is the totality of initial conditions and our program is the laws of nature. These are the whole kit and caboodle: the entire future is already there. No information needs to be added, nothing remains to be discovered. There shall be no novelty, no surprise. Only the clanking of the logical gears remains—a mere formality.

Yet we have learned that in the real world things are always a little messy. Measurements are approximate. Knowledge is imperfect. "The parts have a certain loose play upon one another," said William James, "so that the laying down of one of them does not necessarily determine what the others shall be." James might have been pleasantly surprised by the revelations of quantum physics: the exact states of particles can *never* be perfectly known; uncertainty reigns; probability distributions replace the perfect clockwork dreamt of by Laplace. "It admits that possibilities may be in excess of actualities," James might have said—that is, he did say it, but in advance of the actual science—"and that things not yet revealed to our knowledge may really in themselves be ambiguous." Just so. A physicist with a Geiger counter can never guess when the next click will come. You might think that our modern quantum theorists would join James in cheering indeterminism.

The computers in our thought experiments, if not always the computers we own, are deterministic because people have designed them that way. Likewise, *the laws of science are deterministic because people have written them that way.* They have an ideal perfection that can be attained in the mind or in the Platonic realm but not in the real world. The Schrödinger equation, the screwdriver of modern physics, manages the uncertainties by bundling up the probabilities into a unit, a wave function. It's a ghostly abstract object, this wave function. A physicist can write it as ψ and not worry too much about the contents. "Where did we get that from?" said Richard Feynman. "Nowhere. It's not possible to derive it from anything you know. It came out of the mind of Schrödinger." It just was, and is, astoundingly effective. And once you have it, the Schrödinger equation returns determinism to the process. Calculations are deterministic. Given proper input, good quantum physicists can compute the output with certainty and keep on computing. The only trouble comes in the act of returning from the idealized equations to the real world they are meant to describe. Finally we have to parachute in from the Platonic abstract mathematics to the sublunary stuff on laboratory benches. At that point, when an act of measurement is required, the wave function "collapses," as physicists say. Schrödinger's cat is either alive or dead. According to a limerick:

> It comes as a total surprise
> That what we learn from the ψ's
> Not the fate of the cat
> But related to that:
> The best we can ever surmise.

This collapse of the wave function is the trigger for a special kind of argumentation in quantum physics, not about the mathematics but about the philosophical underpinnings. What can this possibly mean? is the basic problem, and the various approaches are called interpretations. There is the Copenhagen interpretation, first among many. The Copenhagen approach is to treat the collapse of the wave function as an awkward necessity—just a kludge to live with.* The slogan for this interpretation is "Shut up and calculate." There are the Bohmian interpretation, the quantum Bayesian, the objective collapse, and—last but definitely not least—the many worlds. "Go to any meeting, and it is like being in a holy city in great tumult," says the physicist Christopher Fuchs. "You will find all the religions with all their priests pitted in holy war."

The many-worlds interpretation—MWI, to those in the know—is a fantastic piece of make-believe championed by some of the smartest

* Where did this come from, this idea of a "Copenhagen interpretation"? First, "Copenhagen" is cool kids' shorthand for Niels Bohr. For several decades, Copenhagen was to quantum theory what the Vatican is to Catholicism. As for "interpretation," it seems to have started out in German, only the word was *Geist,* as in *Kopenhagener Geist der Quantentheorie* (Werner Heisenberg, 1930).

physicists of our time. They are the intellectual heirs of Hugh Everett, if not Borges. "The MWI is the one with all the glamour and publicity," wrote Philip Ball, the English science writer (ex-physicist), in 2015. "It tells us that we have multiple selves, living other lives in other universes, quite possibly doing all the things that we dream of but will never achieve (or never dare). Who could resist such an idea?" (He can, for one.) The many-worlds champions are like hoarders, unable to throw anything away. There is no such thing as a path not taken. Everything that can happen does happen. All possibilities are realized, if not here, then in another universe. In cosmology universes also abound. Brian Greene has named nine different types of parallel universes: "quilted," "inflationary," "brane," "cyclic," "landscape," "quantum," "holographic," "simulated," and "ultimate." The MWI cannot be demolished by means of logic. It's too appealing: any argument you can make against it has already been considered and (in their minds) refuted by its distinguished advocates.

To me, the most effective physicists are the ones who retain a degree of modesty about their program. Bohr said, "In our description of nature the purpose is not to disclose the real essence of the phenomena but only to track down, so far as it is possible, relations between the manifold aspects of our experience." Feynman said, "I have approximate answers and possible beliefs and different degrees of certainty about different things, but I'm not absolutely sure of anything." Physicists make mathematical models, which are generalizations and simplifications—by definition incomplete, stripped down from the cornucopia of reality. The models expose patterns in the messiness and capitalize upon them. The models themselves are timeless; they exist unchanging. A Cartesian graph plotting time and distance contains its own past and future. The Minkowskian spacetime picture is timeless. The wave function is timeless. These models are ideal, and they are frozen. We can comprehend them within our minds or our computers. The world, on the other hand, remains full of surprises.

William Faulkner said, "The aim of every artist is to arrest motion, which is life, by artificial means and hold it fixed." Scientists do that, too, and sometimes they forget they are using artificial means. You can say Einstein discovered that the universe is a four-dimensional space-time continuum. But it's better to say, more modestly, Einstein discovered that we can describe the universe as a four-dimensional space-time continuum and that such a model enables physicists to calculate almost everything, with astounding exactitude, in certain limited domains. Call it spacetime *for the convenience of reasoning.* Add spacetime to the arsenal of metaphors.

You can say the equations of physics make no distinction between past and future, between forward and backward in time. But if you do, you are averting your gaze from the phenomena dearest to our hearts.* You leave for another day or another department the puzzles of evolution, memory, consciousness, life itself. Elementary processes may be reversible; complex processes are not. In the world of things, time's arrow is always flying.

One twenty-first-century theorist who began to challenge the mainstream block-universe view was Lee Smolin, born in New York in 1955, an expert on quantum gravity and a founder of the Perimeter Institute for Theoretical Physics in Canada. For much of his career

* "That there is a place for the present moment in physics becomes obvious when I take my experience of it as the reality it clearly is to me and recognize that space-time is an abstraction that I construct to organize such experiences," says David Mermin.

Smolin held conventional views of time (for a physicist) before, as he saw it, recanting. "I no longer believe that time is unreal," he declared in 2013. "In fact I have swung to the opposite view: Not only is time real, but nothing we know or experience gets closer to the heart of nature than the reality of time." The rejection of time is itself a conceit. It is a trick that physicists have played on themselves.

"The fact that it is always some moment in our perception, and that we experience that moment as one of a flow of moments, is not an illusion," Smolin wrote. Timelessness, eternity, the four-dimensional spacetime loaf—these are the illusions. Timeless laws of nature are like perfect equilateral triangles. They exist, undeniably, but only in our minds.

> Everything we experience, every thought, impression, intention, is part of a moment. The world is presented to us as a series of moments. We have no choice about this. No choice about which moment we inhabit now, no choice about whether to go forward or back in time. No choice to jump ahead. No choice about the rate of flow of the moments. In this way, time is completely unlike space. One might object by saying that all events also take place in a particular location. But we have a choice about where we move in space. This is not a small distinction; it shapes the whole of our experience.

Determinists, of course, believe that the choice is an illusion. Smolin was willing to treat the persistence of the illusion as a piece of evidence, not to be dismissed glibly, requiring explanation.

For Smolin, the key to salvaging time turns out to be rethinking the very idea of space. Where does that come from? In a universe empty of matter, would space exist? He argues that time is a fundamental property of nature but space is an emergent property. In other words, it is the same kind of abstraction as "temperature": apparent, measurable, but actually a consequence of something deeper and invisible. In the case of temperature, the foundation is the microscopic motion of ensembles of molecules. What we feel as temperature is an average of the energy of these moving molecules. So it is with space: "Space, at the quantum-mechanical level, is not fundamental at all but emergent from a deeper order." (He likewise believes that quantum mechanics itself, with all its puzzles and paradoxes—"cats that are both alive and dead, an infinitude of simultaneously existing universes"—will turn out to be an approximation of a deeper theory.)

For space, the deeper reality is the network of relationships among all the entities that fill it. Things are related to other things; they are connected, and it is the relationships that define space rather than the other way around. This is not a new perspective. It goes back at least to Newton's great rival Leibniz, who refused to accept the view of time and space as containers in which everything is situated—an absolute background for the universe. He preferred to treat them as relations between objects: "Space is nothing else, but That Order or Relation; and is nothing at all without Bodies, but the Possibility of placing them." Empty space is not space at all, Leibniz would say, nor would time exist in an empty universe, because time is the measure of change. "I hold space to be something merely relative, as time is," wrote Leibniz. "Instants, considered without the things, are nothing at all." With the triumph of the Newtonian program, Leibniz's view almost faded from view.

To appreciate the network-centered, relational view of space, we need look no further than the connected, digital world. The internet, like the telegraph a century before, is commonly said to "annihilate" space. It does this by making neighbors of the most distant nodes in a network that transcends physical dimension. Instead of six degrees of separation we have billions of degrees of connectedness. As Smolin put it:

> We live in a world in which technology has trumped the limitations inherent in living in a low-dimensional space. . . . From a cellphone perspective, we live in a 2.5 billion–dimensional space, in which very nearly all our fellow humans are our nearest neighbors. The Internet, of course has done the same thing. The space separating us has been dissolved by a network of connections.

So maybe it's easier now for us to see how things really are. This is what Smolin believes: that time is fundamental but space an illusion; "that the real relationships that form the world are a dynamical network"; and that the network itself, along with everything in it, can and must evolve over time.

He presents a program for further study, based on a notion of "preferred global time" that extends throughout the universe and defines a boundary between past and future. It imagines a family of observers, spread throughout the universe, and a preferred state of rest, against which motion can be measured. Even if "now" need not be the same to different observers, it retains its meaning for the cosmos. These observers, with their persistent sense of a present moment, are a problem to be investigated, rather than set aside.

The universe does what it does. We perceive change, perceive motion, and try to make sense of the teeming, blooming confusion. The hard problem, in other words, is consciousness. We're back where we started, with Wells's Time Traveller, insisting that the only difference between time and space is that "our consciousness moves along it," just before Einstein and Minkowski said the same. Physicists have developed a love-hate relationship with the problem of the self. On the one hand it's none of their business—leave it to the (mere) psychologists. On the other hand, trying to extricate the observer—the measurer, the accumulator of information—from the cool description of nature has turned out to be impossible. Our consciousness is not some magical onlooker; it is a part of the universe it tries to contemplate.

The mind is what we experience most immediately and what does the experiencing. It is subject to the arrow of time. It creates memories as it goes. It models the world and continually compares these models with their predecessors. Whatever consciousness will turn out to be, it's not a moving flashlight illuminating successive slices of the four-dimensional space-time continuum. It is a dynamical system, occurring in time, evolving in time, able to absorb bits of information from the past and process them, and able as well to create anticipation for the future.

Augustine was right all along. The modern philosopher J. R. Lucas, in his *Treatise on Time and Space,* comes back around: "We cannot say what time is, because we know already, and our saying could never match up to all that we already know." So was the Buddha (as translated via Borges): "The man of a past moment has lived, but he does not live nor will he live; the man of a future moment will live, but he has not lived nor does he now live; the man of the present moment lives, but he has not lived nor will he live." We know that the past is gone—it is finished, done, signed, sealed, and delivered. Our access to it is compromised, limited by memories and physical evidence—fossils, paintings in attics, mummies, and old ledgers. We know that eyewitnesses are unreliable and records can be tampered with or misread. The unrecorded past no longer exists. Still, experience persuades us that the past happened and keeps happening. The future is different. The future is yet to come; it is open; not everything can happen but many things can. The world is still under construction.

What is time? Things change, and time is how we keep track.

THIRTEEN

Our Only Boat

Story is our only boat for sailing on the river of time.

—Ursula K. Le Guin (1994)

YOUR NOW IS not my now. You're reading a book. I'm writing a book. You're in my future, yet I know what comes next—some of it—and you don't.*

Then again, you can be a time traveler in your own book. If you're impatient, you can skip ahead to the ending. When memory fails you, just turn back the page. It's all there in writing. You're well acquainted with time traveling by page turning, and so, for that matter, are the characters in your books. "I don't know how to put it exactly," says Aomame in Haruki Murakami's *1Q84,* "but there is a sense of time wavering irregularly when you try to forge ahead. If what is in front is behind, and what is behind is in front, it doesn't really matter, does it?" Soon she appears to be changing her own reality—but you, the reader, can't change history, nor can you change the future. What will be, will be. You are outside it all. You are outside of time.

If this seems a bit meta, it is. In the era of time travel rampant, storytelling has gotten more complicated.

* "Nothing can change the end (written and filed away) of the present chapter," wrote Nabokov partway through *Ada*. Of course, it was not true when he wrote it.

Literature creates its own time. It mimics time. Until the twentieth century, it did that mainly in a sensible, straightforward, linear way. The stories in books usually began at the beginning and ended at the end. A day might pass or many years but usually in order. Time was mostly invisible. Occasionally, though, time came to the foreground. From the beginning of storytelling, there have been stories told inside other stories, and these shift time as well as place: flashbacks and flash-forwards. So aware are we of storytelling that sometimes a character in a story will *feel* like a character in a story, a poor player that struts and frets his hour upon the stage, at time's mercy: *Tomorrow, and tomorrow, and tomorrow* . . . Or perhaps we here in real life develop a nagging suspicion that we are mere characters in someone else's virtual reality. Players performing a script. Rosencrantz and Guildenstern imagine they are masters of their fate, and who are we to know better? The omniscient narrator of Michael Frayn's 2012 novel *Skios* says of the characters living in his story, "If they had been living in a story, they might have guessed that someone somewhere had the rest of the book in his hands, and that what was just about to happen was already there in the printed pages, fixed, unalterable, solidly existent. Not that it would have helped *them* very much, because no one in a story ever knows they are."

In a story one thing comes after another. That is its defining feature. The story is a recital of events. We want to know what happens next. We keep listening, we keep reading, and with any luck the king lets Scheherazade live for one night more. At least this was the traditional view of narrative: "Events arranged in their time sequence," as E. M. Forster said in 1927—"dinner coming after breakfast, Tuesday after Monday, decay after death, and so on." In real life we enjoy a freedom that the storyteller lacks. We lose track of time, we drift and dream. Our past memories pile up, or spontaneously intrude on our thoughts, our expectations for the future float free, but neither memories nor hopes organize themselves into a timeline. "It is always possible for you or me in daily life to deny that time exists and act accordingly even if we become unintelligible and are sent by our fellow citizens to what they choose to call a lunatic asylum," said Forster. "But it is never possible for a novelist to deny time inside the fabric of his novel." In life we may hear the ticking clock or we may not; "whereas in a novel," he said, "there is always a clock."

Not anymore. We have evolved a more advanced time sense—freer and more complex. In a novel there may be multiple clocks, or no clocks, conflicting clocks and unreliable clocks, clocks running backward and clocks spinning aimlessly. "The dimension of time has been shattered," wrote Italo Calvino in 1979; "we cannot love or think except in fragments of time each of which goes off along its own trajectory and immediately disappears. We can rediscover the continuity of time only in the novels of that period when time no longer seemed stopped and did not yet seem to have exploded, a period that lasted no more than a hundred years." He doesn't say exactly when the hundred years ended.

Forster might have known he was oversimplifying, with modernist movements rising self-consciously all around. He had read Emily Brontë, who rebelled against chronological time in *Wuthering Heights*. He had read Laurence Sterne, whose Tristram Shandy had "a hundred difficulties which I have promised to clear up, and a thousand distresses and domestic misadventures crowding in upon me thick and threefold" and threw off the shackles of tense—"A cow broke in (tomorrow morning) to my uncle Toby's fortifications"—and even diagrammed his temporal divagation with a timeline of squiggles, back and forth, up and around.

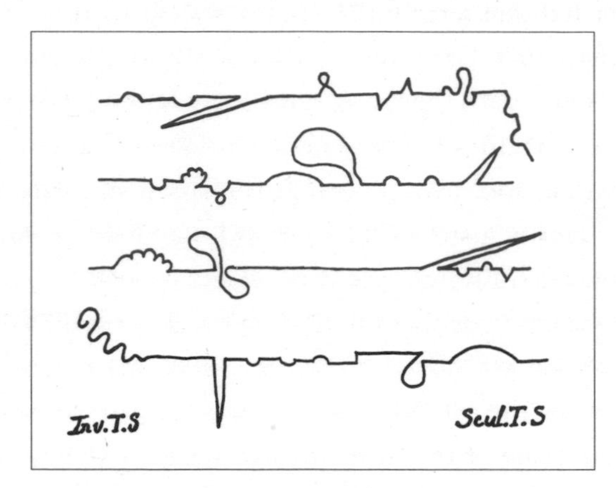

Forster had read Proust, too. But I'm not sure he had gotten the message: that time was busting out all over.

It had seemed that space was our natural dimension: the one we move about in, the one we sense directly. To Proust we became denizens of the time dimension: "I would describe men, even at the risk of giving them the appearance of monstrous beings, as occupying in Time a much greater place than that so sparingly conceded to them in Space, a place indeed extended beyond measure . . . like giants plunged in the years, they touch at once those periods of their lives—separated by so many days—so far apart in Time."* Marcel Proust

* Beckett's translation. (*Si du moins il m'était laissé assez de temps pour accomplir mon oeuvre, je ne manquerais pas de la marquer au sceau de ce Temps dont l'idée s'imposait à moi avec tant de force aujourd'hui, et j'y décrirais les hommes, cela dût-il les faire ressembler à des êtres monstrueux, comme occupant dans le Temps une place*

and H. G. Wells were contemporaries, and while Wells invented time travel by machine, Proust invented a kind of time travel without one. We might call it mental time travel—and meanwhile psychologists have appropriated that term for purposes of their own.

Robert Heinlein's time traveler, Bob Wilson, revisits his past selves—conversing with them and modifying his own life story—and in his way the narrator of *In Search of Lost Time,* sometimes named Marcel, does that, too. Proust, or Marcel, has a suspicion about his existence, perhaps a suspicion of mortality: "that I was not situated somewhere outside of Time, but was subject to its laws, just like the people in novels who, for that reason, used to depress me when I read of their lives, down at Combray, in the fastness of my wicker chair."

"Proust upsets the whole logic of narrative representation," says Gérard Genette, one of the literary theorists who attempted to cope by creating a whole new field of study called narratology. A Russian critic and semioticist, Mikhail Bakhtin, devised the concept of "chronotope" ("time-space," openly borrowed from Einsteinian spacetime) in the 1930s to express the inseparability of the two in literature: the mutual influence they exert upon each other. "Time, as it were, thickens, takes on flesh, becomes artistically visible," he wrote; "likewise, space becomes charged and responsive to the movements of time, plot and history." The difference is that spacetime is just what it is, whereas chronotopes admit as many possibilities as our imaginations allow. One universe may be fatalistic, another may be free. In one, time is linear; in the next, time is a circle, with all our failures,

---

autrement considéable que celle si restreinte qui leur est réservée dans l'espace, une place, au contraire, prolongée sans mesure, puisqu'ils touchent simultanément, comme des géants, plongés dans les années, à des époques vécues par eux, si distantes—entre lesquelles tant de jours sont venus se placer—dans le Temps.)

all our discoveries doomed to be repeated. In one, a man retains his youthful beauty while his picture ages in the attic; in the next, our hero grows backward from senescence to infancy. One story may be ruled by machine time, the next by psychological time. Which time is true? All, or none?

Borges reminds us that Schopenhauer asserted that life and dreams are pages from the same book. To read them in their proper order is to live, but to browse among them is to dream.

The twentieth century gave storytelling a roisterous temporal complexity like nothing that had been seen before. We don't have enough tenses. Or rather, we don't have names for all the tenses we create.* "In what was to have been the future"—that simple clause is the opening of Madeleine Thien's novel *Certainty.* Proust lines a temporal path with mirrors:

> Sometimes passing in front of the hotel he remembered the rainy days when he used to bring his nursemaid that far, on a pilgrimage. But he remembered them without the melancholy that he then

* The problem of verb tense and time travel provides endless fascination in popular culture. Volumes have been written, but most are fictional, beginning with an invention of Douglas Adams in 1980: "The major problem is simply one of grammar, and the main work to consult in this matter is Dr. Dan Streetmentioner's *Time Traveler's Handbook of 1001 Tense Formations.* It will tell you, for instance, how to describe something that was about to happen to you in the past before you avoided it by time-jumping forward two days in order to avoid it. The event will be described differently according to whether you are talking about it from the standpoint of your own natural time, from a time in the further future, or a time in the further past and is further complicated by the possibility of conducting conversations while you are actually traveling from one time to another with the intention of becoming your own mother or father.

"Most readers get as far as the Future Semiconditionally Modified Subinverted Plagal Past Subjunctive Intentional before giving up."

> thought he would surely some day savor on feeling that he no longer loved her. For this melancholy, projected in anticipation prior to the indifference that lay ahead, came from his love. And this love existed no more.

Memories of anticipation, anticipation of memories. To make sense of the time loops narratologists draw symbolic diagrams. We may leave the details to the technicians and savor the new possibilities. *Mixing memory and desire.* The point is that for novelists as much as for physicists the timescape began to replace the landscape. The church of Marcel's childhood is, for him, "an edifice occupying a space with, so to speak, four dimensions—the fourth being Time—extending over the centuries its nave which, from bay to bay, from chapel to chapel, seemed to vanquish and penetrate not only a few yards but epoch after epoch from which it emerged victorious." The other great modernists—especially Joyce and Woolf—likewise made time their canvas and their subject. For all of them, Phyllis Rose has observed, "the prose line wandered in time and space, with any moment in the present acting as a kind of diving platform offering access to a lake of memory, anticipation, and association." Storytelling is unchronological. It is anachronistic. If you are Proust, the narrative of life blends into the life: "life being so unchronological, so anachronistic in its disordering of our days." The narrative itself is the time machine, and memory is the fuel.

Like H. G. Wells, Proust absorbed the new geology. He digs in his own buried strata: "All these memories added to one another now formed a single mass, but one could still distinguish between them—between the oldest, and those that were more recent, born of a fragrance, and then those that were only memories belonging to another person from whom I had learned them—if not fissures, if not true faults, at least that veining, that variegation of coloring, which in certain rocks, in certain marbles, reveal differences in origin, in age, in 'formation.'" We might criticize Proust's view of memory as merely poetic if our modern neuroscientists had settled on a more authoritative model of how memory works, but they have not. Even with the example of computer storage to draw on, even with our detailed neuroanatomies of the hippocampus and the amygdala, no one can really explain how memories are formed and retrieved. Nor can anyone explain away Proust's paradoxical contention: that the past cannot truly be recovered by searching our memories, by interrogating them, by rewinding the film or reaching back into the drawer; rather, that the essence of the past, when it comes to us at all, comes unbidden.

He invented the term "involuntary memory" for this. He warned: "It is a waste of effort for us to try to summon it, all the exertions of our intelligence are useless. The past is hidden outside the realm of our intelligence and beyond its reach." We may think, peering naïvely into our minds, that we have formed our memories and may now call them up for leisurely inspection, but no, the memory we reach for, the memory of the conscious will, is an illusion. "The information it gives about the past preserves nothing of the past itself." Our intelligence rewrites and rewrites again the story it is trying to recall. "The mind feels overtaken by itself; when it, the seeker, is also the obscure country where it must seek." Involuntary memory is the grail for which we may not quest. We don't find it; it finds us. It may lie hidden perchance in a material object—"in the sensation that this material object would give us"—for example, oh, the taste of a *petite madeleine* dipped in lime-blossom tea. It may come in the liminal space between waking and sleep. "Then the confusion among the disordered worlds will be complete, the magic armchair will send him traveling at top speed through time and space."

All things considered, it may seem surprising that it took psychologists sixty more years to define this phenomenon and give it the name "mental time travel," but they have done that now. A neuroscientist in Canada, Endel Tulving, coined the term for what he called "episodic memory" in the 1970s and 1980s. "Remembering, for the rememberer, is mental time travel," he wrote, "a sort of reliving of something that happened in the past." Or the future, naturally. (It's a poor sort of memory that only works backward, remember.) It's MTT, for short, and researchers debate whether it is a uniquely human capability or whether monkeys and birds may also revisit their pasts and project themselves into the future. A more recent definition by two cognitive scientists: "Mental time travel is the ability to mentally project oneself backward in time to relive past experiences and forward in time to pre-live possible future experiences. Previous work has focused on MTT in its voluntary form. Here, we introduce the notion of involuntary MTT." In other words, "involuntary (spontaneous) mental time travel into the past and future." No mention of madeleines, though.

Everyone seems to agree that our imaginations liberate us in the time dimension, even if we can't have a Wellsian time machine. But not Samuel Beckett. The young Dubliner, who had not yet written any of his novels or plays, studied Proust in the summer of 1930 when he was at the École Normale in Paris, in order "to examine in the first place that double-headed monster of damnation and salvation—Time." Freedom is not what he saw. In Proust's world he found only victims and prisoners. Not for Sam "our pernicious and incurable optimism," "our smug will to live," averting our eyes from the bitter fate that lies ahead. We are like organisms of two dimensions, he suggests, like the inhabitants of Flatland, who suddenly discover a third dimension, height. The discovery avails them nothing. They cannot travel in their new dimension. Nor can we. Beckett says:

> There is no escape from the hours and the days. Neither from tomorrow nor from yesterday. There is no escape from yesterday because yesterday has deformed us, or been deformed by us. . . . Yesterday is not a milestone that has been passed, but a drystone on the beaten track of the years, and irremediably part of us, heavy and dangerous.

Beckett will leave to others the pleasures of time travel. For him time is poison. It is a cancer.

> At the best, all that is realized in Time (all Time produce), whether in Art or Life, can only be possessed successively, by a series of partial annexations—and never integrally and at once.

At least he is consistent. We can wait, that's all.

> VLADIMIR: But you say we were here yesterday.
> ESTRAGON: I may be mistaken.

ANY BOOK—bound and sewn, with a beginning, middle, and end—resembles the Universe Rigid. It has a finality lacking in real life, where we can't expect all the threads to tie together when we're done. The novelist Ali Smith says that books are "tangible pieces of time in our hands." You can hold them, you can experience them, but you cannot change them. Except that you can and do: the book is nothing—inert, waiting—until someone is reading it, and then the reader, too, becomes a player in the story. Reading Proust entangles your memories, your desires, with Marcel's. Smith retranslates Heraclitus: "You can't step into the same story twice." Wherever the reader is, on whatever page, the story has a past, which is gone, and a future, which has not yet come.

But surely the reader is capacious, with memory big enough and reliable enough to take in an entire book. (A book is only a few megabytes, after all.) Can't we hold it in the mind all at once—past, present, and future all in our possession? Vladimir Nabokov seemed to think that was the ideal of reading: to possess a book entire, in memory, rather than to encounter it in a state of ignorance or innocence, experiencing it page by page, word by word. "A good reader," said Nabokov in his *Lectures on Literature,* "a major reader, an active and creative reader is a rereader."

> And I shall tell you why. When we read a book for the first time the very process of laboriously moving our eyes from left to right, line after line, page after page, this complicated physical work upon the book, the very process of learning in space and time what the book is about, this stands between us and artistic appreciation.

Ideally a book should be like a painting, which we comprehend (said Nabokov) all at once, outside of time. "When we look at a painting we do not have to move our eyes in a special way even if, as in a book, the picture contains elements of depth and development. The element of time does not really enter in a first contact of a painting."

But can a book really be comprehended whole, all at once, apart from time? Surely a painting is not absorbed in one fell swoop. The eyes roam, the viewer sees this and then that. As for books, they play with time, as music does. They thrive on anticipation, they flirt with expectation. Even if you know a book well—even if you can recite it entire, like the Homeric poet—you cannot experience it as a timeless object. You can appreciate its echoes of memory, its tricks of foreshadowing, but when you read a book you are a creature living in time. The novelist and translator Tim Parks points out the essential role of *forgetting.* "Nabokov does not mention forgetting," he writes, "but it's clear that this is what he is largely talking about." Remember: Memory is not a tape recorder. Not "a stereotype or a tear-sheet." Memories, as Parks says,

> are largely fabrications, re-workings, shifting narratives, simplifications, distortions, photos replacing faces, and so on; what's more, there is no reason to suppose that the original impression is intact somewhere in our heads. We do not possess the past, even that of a few moments ago, and this is hardly a cause for regret, since to do so would severely obstruct our experience of the present.

Also in play is the obverse of forgetting, which is not-yet-knowing. Even the omniscient rereader remembers not-yet-knowing, or where's the fun? No matter how many times we reread a book, we want ignorance of the past, doubt about the future, or we read without expectation, disappointment, suspense, surprise—the panoply of human emotions dependent upon time and forgetting. In Nabokov's *Ada,* someone (the omniscient author or his forgetful narrator) says of his heroes, "Time tricked them, made one of them ask a remembered question, caused the other to give a forgotten answer." They strive "to express something, which until expressed had only a twilight being (or even none at all—nothing but the illusion of the backward shadow of its imminent expression)." Time tricks us all, even fastidious rereaders with time machines.

So even in a book, as in life, closure is an artifice. Someone needs to create it. It is the author who takes on God's job, and as the narratological options grow more convoluted, so do the world-building challenges. "Writing is extremely difficult," says José Saramago, "it is an enormous responsibility, you need only think of the exhausting work involved in setting out events in chronological order, first this one, then that, or, if considered more convenient to achieve the right effect, today's event placed before yesterday's episode, and other no less risky acrobatics, the past treated as if it were new, the present as a continuous process without any present or ending." In turn, readers—and moviegoers—grow ever more aware, learning the tropes and the tricks. We stand on the shoulders of all the time travelers who came before.

Here is a man with a time machine. Perhaps I should say a man *in* a time machine. His name is Charles Yu. He tells us he works in the time-travel industry. He repairs time machines for a living. He's no scientist—just a technician. "To be more specific," he says, "I am a certified network technician for T-Class personal-use chronogrammatical vehicles." For now (troublesome word, in this book) he is living inside one: the TM-31 Recreational Time Travel Device,

> which features an applied temporalinguistics architecture allowing for free-form navigation within a rendered environment, such as, for instance, a story space and, in particular, a science fictional universe.

In other words, we are in a book. It is a story space, a universe. "You get into it. You push some buttons. It takes you to other places, different times. Hit this switch for the past, pull up that lever for the future. You get out and hope the world has changed." Yes, we know all about that by now. We can expect some paradoxes, too.

Charles is a bit of a sad sack. His main companions are a computer UI with a personality skin named TAMMY (sexy software with self-esteem issues) and a "sort of" dog named Ed. The dog was "retconned out of some space western." *Retconning* is a post-post-modern narratological term, short for *retroactive continuity:* after-the-fact rewriting the backstory of a fictional world. Ed doesn't actually exist, though he has a strong smell and licks Charles's face. "Ed is just this weird ontological entity. . . . He must violate some kind of conservation law. Something from nothing: all this saliva." Evidently we should just accept it. Charles does. It's a lonely job: "A lot of people who work in time machine repair are secretly trying to write their novels." Coincidentally, the book we are reading is a first novel by an author named Charles Yu, called *How to Live Safely in a Science Fictional Universe.*

Living in a time machine gives Charles an unusual perspective. Sometimes he feels he exists in a tense: the Present-Indefinite. It's a kind of limbo. It's different from Now. "In any event, what do I need with Now? Now, I think, is overrated. Now hasn't been working out so great for me." Chronological living—everyone just moving forward, looking backward—is yesterday's news. "A kind of lie. That's why I don't do it anymore."

So he sleeps alone, in "a quiet, nameless, dateless day . . . tucked into a hidden cul-de-sac of space-time," and he feels safe there. He has his own mini-wormhole generator that he can use to spy on other universes. Sometimes he has to explain the facts of life to his customers, people who rent time machines in hopes of going back and changing history, or people who rent time machines but *worry* about inadvertently changing history: "Oh God, they say, what if I go back and a butterfly flaps its wings differently and this and that and world war and I never existed and so on and yeah." The rules are that you can't. People never want to hear it, but you can't change the past.

> The universe just doesn't put up with that. We aren't important enough. . . . There are too many factors, too many variables. Time isn't an orderly stream. Time isn't a placid lake recording each of our ripples. Time is viscous. Time is a massive flow. It is a self-healing substance, which is to say, almost everything will be lost.

Charles has learned some more rules. If you ever see yourself coming out of a time machine, run the other way as fast as you can. Nothing good can come of meeting yourself. Try not to have sex with anyone who could possibly be a relative. ("One guy I know ended up as his own sister.") This is twenty-first-century metanarrative: loopy, recursive, self-referential to the *n*th degree. Real science ("real" science) mixes with science-fictional science, which is both a parody of real science and a real science of science fiction. If you see what I mean. Example: "A character within a story, or even a narrator, has, in general, no way of knowing whether or not he is in the past tense narration of a story, or is instead in the present tense (or some other tensed state of affairs) and merely reflecting upon the past."

Above all he misses his father—the father who taught him everything about time travel, who used to say things like, "Today we will journey into Minkowski space," whom he reveres and loves in his memory. So much of time travel is a search for parents, when you think about it. In the *Back to the Future* movies, Marty McFly needs to discover his parents' pasts. His destiny is there. For that matter, the *Terminator* movies are all about finding (killing, protecting) the mother, though the characters don't talk as much about their feelings. "Who wouldn't want to travel back in time and encounter their parents before they become their parents?" asks William Boyd in his 2015 novel *Sweet Caress.* "Before 'mother' and 'father' turned them into figures of domestic myth." We experience childhood one way when we're living it and another way when we relive it in memory. And when we become parents ourselves, we may rediscover our own parents and our own childhoods, as if for the first time. That's the closest we get to having a time machine.

"How can we tell present from past?" Charles's father says this is the key question of time travel. "How do we move the infinitesimal window of the present through the viewfinder at such a constant rate?" It may also be the key question of consciousness. How do we construct the self? Can there be memory without consciousness? Obviously not. Or obviously. It depends what you mean by memory. A rat learns to run a maze—does it remember the maze? If memory is the perpetuation of information, then the least conscious of organisms possess it. So do computers, whose memory we measure in bytes. So does a gravestone. But if memory is the action of recollection, the act of remembrance, then it implies an ability to hold in the mind two constructs, one representing the present and another representing the past, and to compare them, one against the other. How did we learn to distinguish memory from experience? When something misfires and we experience the present as if it were a memory, we call that déjà vu. Considering déjà vu—an illusion or pathology—we might marvel at the ordinary business of remembering.

Can there be consciousness without memory? "We are our memory," said Borges,

> we are that chimerical museum of shifting shapes,
> that pile of broken mirrors.

Our conscious brains invent the concept of time over and over again, inferring it from memory and extrapolating from change. And time is indispensable to our awareness of self. Just as an author does, we construct our own narrative, assemble the scenes in a plausible order, make inferences about cause and effect. Charles's software companion explains, "The book, just like the concept of the 'present,' is a fiction. Which isn't to say it's not real. It's as real as anything else in this science-fictional universe. As real as you are. It's a staircase in a house built by the construction firm of Escher and Sons."

You order the slices of your life. You edit the film even as it records. "Your brain has to trick itself to live in time," she says. Time travel adds a high-octane upgrade to the usual process of creating consciousness.

A HUNDRED YEARS EARLIER, when storytelling seemed simpler and E. M. Forster thought every novel embodied a clock, he invented a story about the future. "Imagine, if you can," he wrote in 1909, "a small room, hexagonal in shape." At its center rests an armchair. In the armchair sits a woman—"a swaddled lump of flesh . . . with a face as white as a fungus." She is happily incarcerated, with every modern comfort:

> There were buttons and switches everywhere—buttons to call for food, for music, for clothing. There was the hot-bath button, by pressure of which a basin of (imitation) marble rose out of the floor, filled to the brim with a warm deodorized liquid. There was the cold-bath button. There was the button that produced literature, and there were of course the buttons by which she communicated with her friends. The room, though it contained nothing, was in touch with all that she cared for in the world.

Most of his contemporaries were still technological optimists and would remain so for another generation, but in his strange novella *The Machine Stops,* Forster creates a grim vision—"a reaction," he admitted later, "to one of the earlier heavens of H. G. Wells." Some unspecified apocalypse, presumably self-inflicted, has driven humanity underground, where people live alone in cells. They have transcended nature and abandoned it. All their needs and desires are met by a global apparatus called the Machine, which is their caretaker and, if they only knew it, their jailer.

> Above her, beneath her, and around her, the Machine hummed eternally; she did not notice the noise, for she had been born with it in her ears. The earth, carrying her, hummed as it sped through silence, turning her now to the invisible sun, now to the invisible stars.

A second apocalypse looms (the title gives it away), but most are oblivious. Just one person sees their imprisonment for what it is. "You know that we have lost the sense of space," he says. "We say 'space is annihilated,' but we have annihilated not space, but the sense thereof. We have lost a part of ourselves."

The "literature epoch" is past. Only one book remains, the Book of the Machine. The Machine is a communications system. It has "nerve-centres." It is decentralized and omnipotent. Humanity worships it. "Through it we speak to one another, through it we see one another, in it we have our being."

Remind you of anything?

FOURTEEN

Presently

We're well past the end of the century when time, for the first time, curved, bent, slipped, flashforwarded and flashbacked yet still kept on rolling along. We know it all now, with our thoughts traveling at the speed of tweet, our 140 characters in search of a paragraph. We're post-history. We're post-mystery.

—Ali Smith (2012)

WHY DO WE NEED time travel, when we already travel through space so far and fast? For history. For mystery. For nostalgia. For hope. To examine our potential and explore our memories. To counter regret for the life we lived, the only life, one dimension, beginning to end.

Wells's *Time Machine* revealed a turning in the road, an alteration in the human relationship with time. New technologies and ideas reinforced one another: the electric telegraph, the steam railroad, the earth science of Lyell and the life science of Darwin, the rise of archeology out of antiquarianism, and the perfection of clocks. When the nineteenth century turned to the twentieth, scientists and philosophers were primed to understand time in a new way. And so were we all. Time travel bloomed in the culture, its loops and twists and paradoxes. We are experts, we are aficionados. Time flies, for us. We know it all now, as Ali Smith says semi-ironically, with our thoughts traveling at the speed of tweet. We are time travelers into our own future. We are Time Lords.

Now another temporal shift has begun, hidden in plain sight.

The people most immersed in the advanced technologies of communication take for granted a persistent connection to others: habitually bearing mobile telephones, flooding the channels with status reports, rumors, factoids. They, we, engage or inhabit a new place, or medium (there is no escaping the awkward terminology). On one hand is the virtual, connected, light-speed realm variously called cyberspace or the internet or the online world or just "the network." On the other hand is everything else, the old place, the "real world." One might say we are living simultaneously in two contrasted forms of society and experience.* Cyberspace is another country. And time? Time happens differently there.

Formerly communication occurred in the present, perforce. You speak, I listen. Your now is my now. Although Einstein showed that the simultaneity was an illusion—signal speed matters, and light takes time to travel from one person's smile to another person's eyes—still, in the main, human intercourse was a melding of present tenses. Then the written word split time: your present became my past, or my future your present. Even a blaze of paint on a cave wall accomplished asynchronous communication. Telephones delivered a new simultaneity—stretching the present across the spatial divide. Voice mail created new opportunities for time shifting. Messaging returns to the instant. And so it continues. The devices, wired and wireless, are always sending and always listening. With persistent connected-

* Marshall McLuhan said that in 1962.

ness time gets tangled. You can't tell the recaps from the prequels. You scrutinize time stamps like tea leaves. The podcast in your earbuds seems more urgent than the ambient voices bleeding through. A river of messages is a "timeline"—*you're in my timeline; I heard it in my timeline*—but the sequence is arbitrary. Temporal ordering can scarcely be trusted. The past, the present, the future go round and collide, bumper cars in a chain of distraction. When distance separates the thunder from the lightning, cyberspace puts them back together.

A DARK AND STORMY NIGHT. A young woman wanders through a boarded-up house snapping photographs. She disregards the posted warning: *Danger Keep Out Unsafe Structure.* Peeling wallpaper reveals letters scrawled on the wall beneath. "Beware . . ." She peels back more paper. "Oh, and duck!" she reads.

"Really, duck!"

"Sally Sparrow, duck, now."

Sally Sparrow (for that is her name) ducks, just in time to avoid a thrown object that smashes the window behind her. Apparently an exercise in asynchronous communication is under way.

This is London, the year 2007, and the writing on the wall is signed "Love from the Doctor (1969)." You, the viewer, know the Doctor as the protagonist of the long-running and multiply reincarnated television series *Doctor Who.* The program had its first go-round on the BBC in 1963, inspired partly by *The Time Machine*—not the book so much as the George Pal movie, released three years before. The Doctor is a survivor of the ancient alien race of Time Lords. He travels through time and space in a vessel called the TARDIS, which for reasons understood only by the most devoted fans has the permanent outward form of a twentieth-century blue British police telephone kiosk. Although the Doctor is an alien from far, far away, with the entire universe at his disposal, his travels are highly Earth centered, and his time-travel adventures favor historical tourism in the style of E. Nesbit's magic amulet and Mr. Peabody's WABAC Machine. He meets Napoleon, Shakespeare, Lincoln, Kublai Khan, Marco Polo, and many English kings and queens. He swaps tradecraft with Einstein. He discovers a time-traveling stowaway called Herbert, whose calling card gives his name as H. G. Wells. Time travel in *Doctor Who* is always good for jokes. Occasionally, however, the problems and paradoxes come to the foreground—never more acutely, never more cleverly than in the story of Sally Sparrow, the episode titled "Blink," written by Steven Moffat and broadcast in 2007.

Still baffled by the writing on the wall, Sally returns to the abandoned house with her friend Kathy Nightingale. Sally loves old things, she says.* We already know that old houses are redolent of time travel. Kathy wanders offscreen. The doorbell rings. Sally answers. A young man hands her a letter from his late grandmother, Kathy Nightingale: *"My dearest Sally Sparrow. If my grandson has done as he promises he will, then as you read these words it has been mere minutes since we last spoke—for you. For me, it has been over sixty years."*

We have a puzzle to solve, we viewers and Sally both. We're getting hints. There are monsters about. Their victims are liable to be transported into the past, willy-nilly, with no way to return.

* "They make me feel sad." What's good about feeling sad? "It's happy for deep people."

If you were trapped in the past, how would you communicate with the future? In a general way, we are all trapped in the past and we are all communicating with the future, via books and epitaphs and time capsules and the rest. But we seldom need to message particular future people at specific future times. A letter for hand delivery by a trusted courier might work, or writing on the wall of an old house. In Terry Gilliam's 1995 movie *Twelve Monkeys* (an elaborate remake of *La jetée*) the unwilling time traveler played by Bruce Willis dials a mysterious telephone number and leaves voice mail. These are one-way messages. Can anyone do better?

Kathy's brother Larry works at a DVD store—that is, he is a specialist in a particular short-lived information medium ("new, second-hand, and rare"). We glimpse television screens in the background. Many of them display the face of one man, whom regular viewers will recognize as none other than the Doctor. Why is he on TV? He seems to be trying to say something urgent. "Don't blink!" for example. He speaks in disconnected fragments. He can be heard explaining in the classic time-traveler tradition: "People don't understand time. It's not what you think it is."

Larry has discovered this man in a hidden track on seventeen different DVDs: "Always hidden away, always a secret," he tells Sally. "It's like he's a ghost DVD extra." Sometimes Larry senses he's hearing one half of a conversation.

The screen starts up again. The Doctor appears to be answering the big question. "People assume that time is a strict progression of cause to effect," he explains, "but actually from a nonlinear, non-subjective viewpoint it's more like a big ball of wibbly wobbly . . . timey wimey . . . stuff."

"*Started* well, that sentence," Sally snarks (for who among us has never talked back to the TV?).

The on-screen Doctor answers, "It got away from me, yeah."

SALLY: Okay, that was weird. Like you can hear me.

THE DOCTOR: Well, I *can* hear you.

Now the conversation begins to get complicated. The Doctor must persuade Sally (and us) that he is a time traveler who has been separated from his time machine (a blue phone box) and hurled back to 1969, that he has been trying to send her messages through an old house and various long-lived human couriers, and that now they are talking to each other via a recording he has concealed on seventeen DVDs, all of which she happens to own in 2007. Larry has heard the Doctor's side of the conversation many times. For him it is preordained: bits laser-engraved on a plastic disc. Finally he is hearing the stereophonic version. Sally talks to the screen, the Doctor talks from the screen, and Larry writes it all down.

SALLY: I've seen this bit before.

THE DOCTOR: Quite possibly.

SALLY: Nineteen sixty-nine, that's where you're talking from?

THE DOCTOR: Afraid so.

SALLY: But you're replying to me. You can't know exactly what I'm going to say, forty years before I say it.

THE DOCTOR [*pedantically*]: Thirty-eight.

How is this possible? Let's review the rules of time travel. Sally is right: he can't hear her. That's an illusion. It's really quite simple, he explains. He possesses a transcript of the entire conversation and is reading his lines, like an actor.*

> SALLY: How can you have a copy of the finished transcript? It's still being written.
>
> THE DOCTOR: I told you. I'm a time traveler. I got it in the future.
>
> SALLY: Okay, let me get my head round this. You're reading aloud from a transcript of a conversation you're still having.
>
> THE DOCTOR: Yeah. Wibbly wobbly, timey wimey.

The TARDIS still needs to reunite with the Doctor. The Doctor still needs to get his hands on the transcript. Before the intricate machinery of this plot is complete, Sally, who now understands the whole story, will have to meet a version of the Doctor who has not yet grasped it. Now her past is his future. "Blink" is all the paradoxes rolled together with a Möbius twist. It's Predestination and Free Will conversing in real time, via technology new for one and obsolete for the other.

By 2007 the internet was in full flow, but it plays no obvious part in the story. Cyberspace is an offstage presence—the dog that doesn't bark in the night. This unusual episode of *Doctor Who* expressed something about our complexified relationship with time. Nowadays, Sally Sparrow's in-box will be overflowing with thousands of emails, mingling past and present, which she may view threaded or flat, and the number only grows, and she is entirely capable of carrying on multiple conversation threads, SMS and MMS, emoji and video, simultaneous and asynchronous, with two participants or many, and meanwhile, with or without earbuds, she hears voices and glimpses

* Like David Tennant, to be exact.

screens everywhere, in waiting rooms and on signposts, and if she pauses to think, she may have trouble placing all the information in proper temporal sequence—wibbly wobbly, timey wimey—but who pauses to think?

WHEN THE BROTHERS Louis and Auguste Lumière invented the *cinématographe* in the 1890s, they did not begin by filming actors dressed in costumes. They did not make fictional movies. They trained operators in the new technology and sent Clément and Constant and Félix and Gaston and many more across the globe to record snippets of real life. Naturally they filmed workers leaving their own factory—who could resist *La sortie de l'usine Lumière à Lyon*?—but by 1900 they were filming a cockfight in Guadalajara, and the foot traffic on Broadway, and men smoking opium in what is now Vietnam. Audiences flocked to see these scenes of faraway live action. The creation of these images marks an event horizon. When we look back, the pre-1900 past is less visible. It's good we have books.

So much of the world comes to us on screens now, with sound as lifelike as the picture. The screens range farther than anyone could ever see unaided. Who is to say that these are not time gates? People "stream" music to us and video, the tennis match we're watching may or may not be "live," the people in the stadium watching the instant replay on the stadium screen, which we see repeated on our screen, may have done that yesterday, in a different time zone. Politicians record their responses to speeches they have not yet seen, for instant broadcast. If we confuse the real world with our many virtual worlds, it's because so much of the real world is virtual. For many people, there is no personal memory of a time without screens. So many windows, so many clocks.

"Internet time" became a term of art. Andrew Grove, chief executive of Intel, 1996: "We are now living on internet time." Often this was just a cool-kids way of saying "faster," but our relationship to time was changing yet again, even if no one quite understood what or how. On internet time the past bleeds into the present. And the future? There seems to be a feeling that the future is already here. Blink and it has happened. Thus the future vanishes.

"Increasingly, our concepts of past, present and future are being forced to revise themselves," wrote J. G. Ballard in 1995—science fiction, as ever, the canary in the coal mine. "The future is ceasing to exist, devoured by the all-voracious present. We have annexed the future into the present, as merely one of those manifold alternatives open to us."

We are annexing the past as well. Institutions from *Scientific American* to *The Bridge World* spill open their archives to reveal what was new *50 Years Ago*. The online front page of the *New York Times* recycles its first reporting on bagels and pizza. Backward reels the global mind. Just when the obsession with newness seemed more ferocious than ever, Svetlana Boym, a time-twisting theorist of nostalgia, observed: "The first decade of the twenty-first century is not characterized by the search for newness, but by the proliferation of nostalgias that are often at odds with one another. Nostalgic cyberpunks and nostalgic hippies, nostalgic nationalists and nostalgic cosmopolitans, nostalgic environmentalists and nostalgic metrophiliacs (city lovers) exchange pixel fire in the blogosphere." For all this blooming shape-shifting nostalgia we can thank the time travelers. "The object of romantic nostalgia must be beyond the present space of experience," Boym writes, "somewhere in the twilight of the past or on the island of utopia where time has happily stopped, as on an antique clock."

What a strange ending for the twentieth century! The new century—the new millennium, for those who were counting—arrived with televised fireworks and bands playing (and computer panic) but scarcely a glimmer of the glorious optimism that lit up the year 1900, when everyone seemed to be rushing to the prow of a great ship and gazing hopefully toward the horizon, dreaming of their scientific future: airships, moving sidewalks, *Schönwettermaschinen*, underwater croquet, flying cars, gas-powered cars, flying people. *Andiamo, amici!* Many of those dreams came true. So when the new millennium dawned, what bright dreams for the year 3000? Or the year 2100?

Newspapers and websites polled their readers for predictions. They were disappointing. *We will control the weather.* (Again.) *Deserts will become tropical forests.* Or the reverse. *Space elevators.* But not much

Card produced c. 1900 by Hildebrands chocolate company

space travel. Warp drive and wormholes notwithstanding, we seem to have given up on populating the galaxy. *Nanorobots. Remote-control warfare.* The internet in your contact lens or brain implant. Self-driving cars, a comedown, somehow, from *i futuristi* and their fearsome roaring racing machines. The aesthetic of futurism changed, too, without anyone issuing a manifesto—from big and bold, primary colors and metallic shine to grim, dank rot and ruins. Genetic engineering and/or species extinctions. Is that all the future we have to look forward to? Nanobots and self-driving cars?

If we lack space travel, we do have telepresence. "Present" in this context pertains to space, not to time. Telepresence was born in the 1980s, when remotely controlled cameras and microphones came into their own. Deep sea explorers and bomb squads can project themselves elsewhere—project their souls, their eyes and ears, while the body remains behind. We send robots beyond the planets and inhabit them. In the same decade the word *virtual,* already by then a computer term, began to refer to remote simulations—virtual office, virtual town halls, virtual sex. And, of course, virtual reality. Another way to look at telepresence is that people virtualize themselves.

A women finds herself piloting a quadcopter in a slightly creepy "beta of some game"—like a first-person shooter with "nothing to shoot"—and because she is a character in a novel by William Gibson (*The Peripheral,* 2014) we must already wonder what is virtual and what is real. Her name is Flynne and she seems to live somewhere in the American South—back country, trailer down by the creek. But in the present or the future? Hard to know exactly. At the very least, waves of the future are lapping at the shore. Marine vets have scars, physical and mental, from implanted "haptics." The era's namespace includes Cronut, Tesla, Roomba, Sushi Barn, and Hefty Mart.

Roadside storefronts offer "fabbing"—three-dimensional printing of practically everything. Drones are ascendant. Every buzzing insect is a potential spy.

Anyway, Flynne leaves her reality behind to pilot her drone through a different, virtual reality. A mysterious (virtual?) corporate entity is paying her to do it. She hovers near a great dark building. She looks up—camera up. She looks down—camera down. "All around her were whispers, urgent as they were faint, like a cloud of invisible fairy police dispatchers." Everyone knows how immersive a computer game can become, but what is her goal? Her purpose? Apparently she is meant to chase away other drones, which swarm like dragonflies, but it doesn't feel like any game she has played before.* Then—a window, a woman, a balcony—Flynne witnesses a murder.

We have met Gibson before: the futurist who denies writing about the future. It was Gibson who invented the word *cyberspace* in 1982 after watching kids playing video games at an arcade in Vancouver, staring into their consoles, turning knobs and pounding buttons to manipulate a universe no one else could see. "It seemed to me that what they wanted was to be inside the games, within the notional space of the machine," he said later. "The real world had disappeared for them—it had completely lost its importance. They were in that notional space." There was no such thing as cyberspace then—as Gibson imagined it, "a consensual hallucination experienced daily by billions of legitimate operators, in every nation." The space behind all the computers. "Lines of light ranged in the nonspace of the mind, clusters and constellations of data." We all feel that way sometimes.

* "Feels more like working security than a game."
"Maybe it's a game about working security."

At some point it occurred to Gibson that he had been describing something like the "Aleph" of Borges's 1945 story: a point in space that contains all other points. To see the Aleph you must lie flat and immobile in darkness. "A certain ocular adjustment will also be necessary." What you see then cannot be contained in words, Borges writes,

> for any listing of an endless series is doomed to be infinitesimal. In that single gigantic instant I saw millions of acts both delightful and awful; not one of them occupied the same point in space, without overlapping or transparency. What my eyes beheld was simultaneous, but what I shall now write down will be successive, because language is successive.

The *space* in cyberspace vanishes. It collapses into a network of connections: as Lee Smolin said, a billion-dimensional space. Interaction is all. And what of cybertime? Every hyperlink is a time gate.* Millions of acts both delightful and awful—posts, tweets, comments, emails, "likes," swipes, winks—appear simultaneously or successively. Signal speed is light speed, time zones overlap, and time stamps shift like motes in a sunbeam. The virtual world is build on transtemporality.

Gibson, who always felt time travel to be an implausible magic, avoided it through ten novels written across thirty years.† Indeed, as

* "—Must be a spatio-temporal hyperlink."
"—What's that?"
"—No idea. Just made it up. Didn't want to say 'magic door.'"
—Steven Moffat, "The Girl in the Fireplace" (*Doctor Who*), 2006

† Completists will note, however, his 1981 story "The Gernsback Continuum," a hat tip to Hugo. The story is at least time travelish. Semiotic ghosts. "As I moved among these secret ruins, I found myself wondering what the inhabitants of that lost future would think of the world I lived in."

his imagined futures kept crowding in on the conveyor belt of the present, he renounced the future altogether. "Fully imagined futures were the luxury of another day, one in which 'now' was of some greater duration," says Hubertus Bigend in the 2003 *Pattern Recognition.* "We have no future because our present is too volatile." The future stands upon the present, and the present is quicksand.

Back to the future once more, though, in Gibson's eleventh novel, *The Peripheral.* A near future interacts with a far future. Cyberspace gave him a way in. New rules of time travel: matter cannot escape its time but information can. The future discovers that it can *email* the past. Then it *phones* the past. The information flows both ways. Instructions are sent for 3-D fabbing: helmets, goggles, joysticks. It is a marriage of time shifting and telepresence.

To the people of the future, the denizens of the past can be employed as "polts" (from *poltergeist*—"ghosts that move things, I suppose"). Money can be sent or created (win lotteries, manipulate the stock market). Finance has become virtual, after all. Corporations are shells, built of documents and bank accounts. It's outsourcing in a new dimension. Does the manipulation of people across time create headaches? "Far less than the sort of paradox we're accustomed to culturally, in discussing imaginary transtemporal affairs. It's actually quite simple." After all, we know about time forks. We are aficionados of branching universes. "The act of connection produces a fork in causality, the new branch causally unique. A stub, as we call them."

Not that paradoxes are unknown. At one point a future law-enforcement agent called Detective Inspector Ainsley Lowbeer explains to an avatar—exoskeleton, homunculus, *peripheral*—inhabited by Flynne, "I'm told that arranging your death would in no way constitute a crime here, as you are, according to current best legal opinion, not considered to be real." Nanobots are real. Cosplay is real. Drones are real. Futurity is done.

WHY DO WE NEED time travel? All the answers come down to one. To elude death.

Time is a killer. Everyone knows that. Time will bury us. *I wasted time, and now doth time waste me.* Time makes dust of all things. Time's winged chariot isn't taking us anywhere good.

How aptly named, the time beyond death: the Hereafter. The past, in which we did not exist, is bearable, but the future, in which we will not exist, troubles us more. I know that in the vast expanse of space I am an infinitesimal mote—fine. But confinement to an eyeblink of time, an instant never to return, is harder to accept. Of course, before inventing time travel, human cultures found other ways to soften the unpleasantness. One may believe in the soul's immortality, in cycles of transmigration and reincarnation, in a paradisical afterlife. The time capsulists, too, are preparing transport to the afterlife. Science provides cold comfort—as Nabokov says, "problems of space and time, space versus time, time-twisted space, space as time, time as space—and space breaking away from time, in the final tragic triumph of human cogitation: I am because I die."* Time travel at least sets our imaginations free.

Intimations of immortality. Maybe that's the best we can hope for. What is the fate of Wells's Time Traveller? For his friends he is gone but perhaps not dead. "He may even now—if I may use the

* Heidegger: "We perceive time only because we know we have to die."

phrase—be wandering on some plesiosaurus-haunted Oolitic coral reef, or beside the lonely saline lakes of the Triassic Age." Entropy can be held off only here and there, now and then. Every life lapses into oblivion. *Time and the bell have buried the day.* Einstein was explicit about seeking solace in the spacetime view ("Now he has left this strange world a little before me. This means nothing"), and so likewise is Kurt Vonnegut's narrator in *Slaughterhouse-Five:*

> The most important thing I learned on Tralfamadore was that when a person dies he only appears to die. He is still very much alive in the past, so it is very silly for people to cry at his funeral. . . . It is just an illusion we have here on Earth that one moment follows another one, like beads on a string, and that once a moment is gone it is gone forever. When a Tralfamadorian sees a corpse, all he thinks is that the dead person is in a bad condition in that particular moment, but that the same person is just fine in plenty of other moments.

Some comfort there. You lived; you will always have lived. Death does not erase your life. It is mere punctuation. If only time could be seen whole, then you could see the past remaining intact, instead of vanishing in the rearview mirror. There is your immortality. Frozen in amber.

For me the price of denying death in this way is denying life. *Dive back into the flux. Turn your face toward sensation, that flesh-bound thing.*

> By this, and this only, we have existed
> Which is not to be found in our obituaries

Or in memories draped by the beneficent spider
Or under seals broken by the lean solicitor
In our empty rooms

Every death is an obliteration of memory. To counter, the online world promises a collective, connected memory and thus offers an ersatz immortality. In cyberspace, the present moment churns and past moments aggregate. @SamuelPepys, tweeting his diary day by day, is one of "ten dead people" the *Telegraph* (London) recommends we follow, because "Twitter isn't solely the preserve of living beings." Facebook announced procedures for continuing or "memorializing" the accounts of its deceased customers. A startup called Eter9 offered to "externalize" (*and* "eternalize") customers in the persons of artificial agents. Evidently corporeal death is no reason to stop posting and commenting: "The Counterpart is your Virtual Self that will stay in the system and interact with the world just like you would if you were present." No wonder science-fiction writers despair of inventing the future. Eternity isn't what it used to be. Heaven was better in the good old days. Peering toward the afterlife, we can look forward and we can look back.

"When I look back all is flux," writes John Banville, "without beginning and flowing towards no end, or none that I shall experience, except as a final full stop."

What comes next? After the final full stop, nothing. After the modern—the postmodern, of course. The avant-garde. Futurism. You can read about all these epochs in the history books of the prewired world. Ah, the good old days.

When the future vanishes into the past so quickly, what remains is a kind of atemporality, a present tense in which temporal order feels as arbitrary as alphabetical order. We say that the present is real—yet it flows through our fingers like quicksilver. It slips away: *now*— no, *now*— wait, *now* . . . Psychologists try to measure the length of *now* as felt in, or perceived by, the brain. It's hard to know just what to measure. Two sounds as close together as a millisecond tend to be perceived as one. Two flashes of light seem simultaneous even when they are one-hundredth of a second apart. Even when we recognize separate stimuli, we can't reliably say which came first until they are close to a tenth of a second apart. Psychologists suggest that what we call *now* is a rolling period of two or three seconds. William James's term was the "specious present": this illusion, he said, "varying in length from a few seconds to probably not more than a minute . . . is the original intuition of time." Borges had his own intuitions: "They tell me that the present, the 'specious present' of the psychologists, lasts between several seconds and the smallest fraction of a second, which is also how long the history of the universe lasts. Or better, there is no such thing as 'the life of a man,' nor even 'one night in his life.' Each moment we live exists, not the imaginary combination of these moments." Immediate sensation dissolves into short-term memory.

In the wired world, creating the present becomes a communal process. Everyone's mosaic is crowd-sourced, a photomontage with multiple perspectives. Images of the past, fantasies of the future, live videocams, all shuffled and blended. All time and no time. The path back through history is cluttered, the path forward cloudy. "Fare forward, travellers!" Eliot said, "not escaping from the past/Into different lives, or into any future." Without the past for background and frame, the present is only a blur. "Where is it, this present?" asked James. "It has melted in our grasp, fled ere we could touch it, gone in the instant of becoming." The brain has to assemble its putative present from a hodgepodge of sensory data, continually compared and contrasted with a succession of previous instants. It might be fair to say that all we perceive is change—that any sense of stasis is a constructed illusion. Every moment alters what came before. We reach across layers of time for the memories of our memories.

"Live in the now," certain sages advise. They mean: focus; immerse yourself in your sensory experience; bask in the incoming sunshine, without the shadows of regret or expectation. But why should we toss away our hard-won insight into time's possibilities and paradoxes? We lose ourselves that way. "What more terrifying revelation can there be than that it is the present moment?" wrote Virginia Woolf. "That we survive the shock at all is only possible because the past shelters us on one side, the future on another." Our entry into the past and the future, fitful and fleeting though it may be, makes us human.

So we share the present with ghosts. An Englishman builds a machine in guttering lamplight, a Yankee engineer awakens in medieval fields, a jaded Pennsylvania weatherman relives a single February day, a little cake summons lost time, a magic amulet transports schoolchildren to golden Babylon, torn wallpaper reveals a timely message, a boy in a DeLorean seeks his parents, a woman on a pier awaits her lover—all these, our muses, our guides, in the unending now.

Acknowledgments

For pointers and discussion I am deeply obliged to David Albert, Lera Boroditsky, Billy Collins, Uta Frith, Chris Fuchs, Rivka Galchen, William Gibson, Janna Levin, Alison Lurie, Daniel Menaker, Maria Popova, Robert D. Richardson, Phyllis Rose, Siobhan Roberts, Lee Smolin, Craig Townsend, and Grant Wythoff, as well as my indefatigable agent, Michael Carlisle, my wise and patient editor, Dan Frank, and, always, Cynthia Crossen.

Sources and Further Reading

These are some of the works on which this one depends.

STORIES

Edwin Abbott Abbott, *Flatland*, 1884.

Douglas Adams, "The Pirate Planet" (*Doctor Who*), 1978.
The Restaurant at the End of the Universe, 1980.

Woody Allen, *Sleeper*, 1973.
Midnight in Paris, 2011.

Kingsley Amis, *The Alteration*, 1976.

Martin Amis, "The Time Disease," 1987.
Time's Arrow, 1991.

Isaac Asimov, *The End of Eternity*, 1955.

John Jacob Astor IV, *A Journey in Other Worlds*, 1894.

Kate Atkinson, *Life After Life*, 2013.
A God in Ruins, 2014.

Marcel Aymé, "*Le décret*," 1943.

John Banville, *The Infinities*, 2009.
Ancient Light, 2012.

Max Beerbohm, "Enoch Soames," 1916.

Edward Bellamy, *Looking Backward,* 1888.

Alfred Bester, "The Men Who Murdered Mohammed," 1958.

Michael Bishop, *No Enemy but Time,* 1982.

Jorge Luis Borges, *El jardín de senderos que se bifurcan,* 1941.
El aleph, 1945.
Nueva refutación del tiempo, 1947.

Ray Bradbury, "A Sound of Thunder," 1952.

Ted Chiang, "Story of Your Life," 1998.

Ray Cummings, *The Girl in the Golden Atom,* 1922.

Philip K. Dick, *The Man in the High Castle,* 1962.
Counter-Clock World, 1967.
"A Little Something for Us Tempunauts," 1974.

Daphne du Maurier, *The House on the Strand,* 1969.

T. S. Eliot, *Four Quartets,* 1943.

Harlan Ellison, "The City on the Edge of Forever" (*Star Trek*), 1967.

Ralph Milne Farley, "I Killed Hitler," 1941.

Jack Finney, "The Face in the Photo," 1962.
Time and Again, 1970.

F. Scott Fitzgerald, "The Curious Case of Benjamin Button," 1922.

E. M. Forster, *The Machine Stops,* 1909.

Stephen Fry, *Making History,* 1997.

Rivka Galchen, "The Region of Unlikeness," 2008.

Hugo Gernsback, *Ralph 124C 41+: A Romance of the Year 2660,* 1925.

David Gerrold, *The Man Who Folded Himself,* 1973.

William Gibson, "The Gernsback Continuum," 1981.
The Peripheral, 2014.

Terry Gilliam, *Twelve Monkeys,* 1995.

James E. Gunn, "The Reason Is with Us," 1958.

Robert Harris, *Fatherland,* 1992.

Robert Heinlein, "Life-Line," 1939.
"By His Bootstraps," 1941.
Time for the Stars, 1956.
" '—All You Zombies—,' " 1959.

Washington Irving, "Rip Van Winkle," 1819.

Henry James, *The Sense of the Past,* 1917.

Alfred Jarry, *"Commentaire pour servir à la construction pratique de la machine à explorer le temps,"* 1899.

Rian Johnson, *Looper,* 2012.

Ursula K. Le Guin, *The Lathe of Heaven,* 1971.
A Fisherman of the Inland Sea, 1994.

Muray Leinster (William Fitzgerald Jenkins), "The Runaway Skyscraper," 1919.

Stanisław Lem, *Memoirs Found in a Bathtub,* 1961.
The Futurological Congress, 1971.

Alan Lightman, *Einstein's Dreams,* 1992.

Samuel Madden, *Memoirs of the Twentieth Century,* 1733.

Chris Marker, *La jetée,* 1962.

J. McCullough, *Golf in the Year 2000; or, What Are We Coming To,* 1892.

Louis-Sébastien Mercier, *L'an deux mille quatre cent quarante: rêve s'il en fût jamais,* 1771.

Edward Page Mitchell, "The Clock That Went Backward," 1881.

Steven Moffat, "Blink" (*Doctor Who*), 2007.

Vladimir Nabokov, *Ada, or Ardor,* 1969.

Edith Nesbit, *The Story of the Amulet,* 1906.

Audrey Niffenegger, *The Time Traveler's Wife,* 2003.

Dexter Palmer, *Version Control,* 2016.

Edgar Allan Poe, "The Power of Words," 1845.
"Mellonta Tauta: On Board Balloon 'Skylark,' April 1, 2848," 1849.

Marcel Proust, *À la recherche du temps perdu,* 1913–27.

Harold Ramis and Danny Rubin, *Groundhog Day,* 1993.

Philip Roth, *The Plot Against America,* 2004.

W. G. Sebald, *Austerlitz,* 2001.

Clifford D. Simak, *Time and Again,* 1951.

Ali Smith, *How to Be Both,* 2014.

George Steiner, *The Portage to Cristóbal of A.H.,* 1981.

Tom Stoppard, *Arcadia,* 1993.

William Tenn, "Brooklyn Project," 1948.

Mark Twain (Samuel Clemens), *A Connecticut Yankee in King Arthur's Court,* 1889.

Jules Verne, *Paris au XXe siècle,* 1863.

Kurt Vonnegut, *Slaughterhouse-Five,* 1969.

H. G. Wells, *The Time Machine,* 1895.
The Sleeper Awakes, 1910.

Connie Willis, *Doomsday Book,* 1992.

Virginia Woolf, *Orlando,* 1928.

Charles Yu, *How to Live Safely in a Science Fictional Universe,* 2010.

Robert Zemeckis and Bob Gale, *Back to the Future,* 1985.

ANTHOLOGIES

Mike Ashley, *The Mammoth Book of Time Travel SF,* 2013.

Peter Haining, *Timescapes,* 1997.

Robert Silverberg, *Voyagers in Time,* 1967.

Harry Turtledove and Martin H. Greenberg, *The Best Time Travel Stories of the Twentieth Century,* 2004.

Ann and Jeff Vandermeer, *The Time Traveler's Almanac,* 2013.

BOOKS ABOUT TIME TRAVEL AND TIME

Paul E. Alkon, *Origins of Futuristic Fiction,* 1987.

Kingsley Amis, *New Maps of Hell,* 1960.

Isaac Asimov, *Futuredays,* 1986.

Anthony Aveni, *Empires of Time,* 1989.

Svetlana Boym, *The Future of Nostalgia,* 2001.

Jimena Canales, *The Physicist and the Philosopher,* 2015.

Sean Carroll, *From Eternity to Here,* 2010.

Istvan Csicsery-Ronay, Jr., *The Seven Beauties of Science Fiction,* 2008.

Paul Davies, *About Time,* 1995.
How to Build a Time Machine, 2001.

John William Dunne, *An Experiment with Time,* 1927.

Arthur Eddington, *The Nature of the Physical World,* 1928.

J. T. Fraser, ed., *The Voices of Time,* 1966, 1981.

Peter Galison, *Einstein's Clocks, Poincaré's Maps. Empires of Time,* 2004.

J. Alexander Gunn, *The Problem of Time,* 1929.

Claudia Hammond, *Time Warped,* 2013.

Diane Owen Hughes and Thomas R. Trautmann, eds., *Time: Histories and Ethnologies,* 1995.

Robin Le Poidevin, *Travels in Four Dimensions,* 2003.

Wyndham Lewis, *Time and Western Man,* 1928.

Michael Lockwood, *The Labyrinth of Time,* 2005.

J. R. Lucas, *A Treatise on Time and Space,* 1973.

John W. Macvey, *Time Travel,* 1990.

Paul J. Nahin, *Time Machines,* 1993.

Charles Nordmann, *The Tyranny of Time* (*Notre maître le temps*), 1924.

Clifford A. Pickover, *Time: A Traveler's Guide,* 1998.

Paul Ricoeur, *Time and Narrative* (*Temps et récit*), 1984.

Lee Smolin, *Time Reborn,* 2014.

Stephen Toulmin and June Goodfield, *The Discovery of Time,* 1965.

Roberto Mangabeira Unger and Lee Smolin, *The Singular Universe and the Reality of Time,* 2014.

David Foster Wallace, *Fate, Time, and Language,* 2010.

Gary Westfahl, George Slusser, and David Leiby, eds., *Worlds Enough and Time,* 2002.

David Wittenberg, *Time Travel: The Popular Philosophy of Narrative,* 2013.

Index

Illustration Credits

Page 11: From *The Dublin Review*, January–June 1920, vol. 166. Courtesy of Stanford University Library.

Page 14: Courtesy of the New York Public Library.

Page 22: Still image from episode 41 of *Rocky & Bullwinkle & Friends*, copyright © 2004 by DreamWorks Animation LLC. Used by permission.

Page 32: From *A Connecticut Yankee in King Arthur's Court* by Mark Twain. New York: Charles L. Webster & Co., 1889.

Page 38: From Wikimedia Commons.

Page 61: Still image from *Felix the Cat Trifles with Time*, copyright © DreamWorks Animation LLC. Used by permission.

Page 68: From *Science and Invention in Pictures*, July 1925.

Page 95: Courtesy of the Robert A. and Virginia Heinlein Archives and the Heinlein Prize Trust.

Page 170: From *The Story of the Westinghouse Time Capsule*. East Pittsburgh, Penn.: Westinghouse Electric & Manufacturing Company, 1938.

Page 180 (*top*): From *The Book of Record of the Time Capsule of Cupaloy,* New York World's Fair, 1939. New York: Westinghouse Electric & Manufacturing Company, 1938.

Page 180 (*bottom*): From *The Book of Record of the Time Capsule of Cupaloy,* New York World's Fair, 1939. New York: Westinghouse Electric & Manufacturing Company, 1938.

Page 192: From *E. Nesbit: A Biography* by Doris Langley Moore. Philadelphia: Chilton Company, 1966.

Page 243: Still image from *La Jetée* by Chris Marker, copyright © 1963 Argos Films.

Page 276: From *The Life and Opinions of Tristram Shandy, Gentleman* by Laurence Sterne, Chapter XXXVIII.

Page 304: Courtesy of South West News Service Ltd.

ABOUT THE AUTHOR

JAMES GLEICK (www.around.com) was born in New York City in 1954. He graduated from Harvard College and worked at the *New York Times* for ten years as a reporter and editor. His six previous books include *Chaos* (1987) and *The Information* (2011), as well as biographies of Isaac Newton and Richard Feynman. They have been translated into thirty languages.

A NOTE ON THE TYPE

This book was set in Granjon, a type named in compliment to Robert Granjon, a type cutter and printer active in Antwerp, Lyons, Rome, and Paris from 1523 to 1590. Granjon, the boldest and most original designer of his time, was one of the first to practice the trade of typefounder apart from that of printer.

Linotype Granjon was designed by George W. Jones, who based his drawings on a face used by Claude Garamond (ca. 1480–1561) in his beautiful French books. Granjon more closely resembles Garamond's own type than do any of the various modern faces that bear his name.

Composed by North Market Street Graphics,
Lancaster, Pennsylvania